T0207432

Lecture Notes in Mathematics

Volume 2323

This series reports on new developments in all areas of mathematics and their applications - quickly, informally and at a high level. Mathematical texts analysing new developments in modelling and numerical simulation are welcome. The type of material considered for publication includes:

1. Research monographs
2. Lectures on a new field or presentations of a new angle in a classical field
3. Summer schools and intensive courses on topics of current research.

Texts which are out of print but still in demand may also be considered if they fall within these categories. The timeliness of a manuscript is sometimes more important than its form, which may be preliminary or tentative.

Titles from this series are indexed by Scopus, Web of Science, Mathematical Reviews, and zbMATH.

Alan Carey • Galina Levitina

Index Theory Beyond the Fredholm Case

Alan Carey
School of Mathematics and Applied Statistics
University of Wollongong
Wollongong, NSW, Australia

Galina Levitina
Mathematical Sciences Institute
Australian National University
Canberra, ACT, Australia

ISSN 0075-8434 ISSN 1617-9692 (electronic)
Lecture Notes in Mathematics
ISBN 978-3-031-19435-1 ISBN 978-3-031-19436-8 (eBook)
https://doi.org/10.1007/978-3-031-19436-8

This Springer imprint is published by the registered company Springer Nature Switzerland AG
The registered company address is: Gewerbestrasse 11, 6330 Cham, Switzerland

Preface

These notes are about extending index theory to some examples where non-Fredholm operators arise. They are far from comprehensive. We have focussed on one aspect of the problem of what replaces the notion of spectral flow and the Fredholm index when the operators in question have zero in their essential spectrum. Most work in this topic stems from the so-called Witten index that is discussed at length here. The new direction described in these notes is the introduction of 'spectral flow beyond the Fredholm case'.

Spectral flow was introduced by Atiyah-Patodi-Singer [APS76] for elliptic operators on odd dimensional compact manifolds. They argued that it could be computed from the Fredholm index of an elliptic operator on a manifold of one higher dimension. A general proof of this fact was produced by Robbin-Salamon [RS95] but was restricted to operators with discrete spectrum as occur in the study of elliptic operators on compact manifolds.

In [GLM$^+$11], a start was made on extending the Robbin-Salamon theorem to operators with some essential spectrum as occurs on non-compact manifolds. The new ingredient introduced there was to exploit scattering theory following the fundamental paper [Pus08]. These results do not apply to differential operators directly, only to pseudo-differential operators on manifolds, due to the restrictive assumption that spectral flow is only considered between an operator and its perturbation by a relatively trace-class operator.

In these lecture notes, we give an expository account of the main results of these earlier papers and their generalisation to the study of spectral flow between an operator and a perturbation satisfying a higher pth Schatten class condition for $0 \leq p < \infty$, thus allowing differential operators on manifolds of any dimension $d < p + 1$. To achieve this, we establish an operator trace formula that does not assume any ellipticity or Fredholm properties at all. This operator trace formula is motivated by Benameur et al. [BCP$^+$06] and Carey et al. [CGK16]. We then explain some applications to index theory and spectral flow. Examples may be obtained by using Dirac type operators on $L_2(\mathbb{R}^d)$ for arbitrary $d \in \mathbb{N}$ (see Sect. 7.1). In this setting, Theorem 6.2.2 substantially extends [CGG$^+$16, Theorem 3.5], where the case $d = 1$ was treated.

We now briefly explain the central point of these notes using notation listed in the Notations. Our discussion focuses on a family $\{B(t)\}_{t\in\mathbb{R}}$ of bounded self-adjoint operators on the Hilbert space $\mathcal{H}$ that are p-relative trace-class perturbations of an unbounded self-adjoint operator A_- (see Hypothesis 3.2.5 for the precise assumptions). Denote by B_+ the uniform norm asymptote as $t \to \infty$ of the family $\{B(t)\}_{t\in\mathbb{R}}$. Introduce the model operator $\boldsymbol{D_A}$ in $L_2(\mathbb{R};\mathcal{H})$ by first defining $\boldsymbol{A}$ in $L_2(\mathbb{R};\mathcal{H})$ as $(\boldsymbol{A}f)(t) = (A_- + B(t))f(t)$ and then setting

$$\boldsymbol{D_A} = \frac{d}{dt} + \boldsymbol{A}, \quad \operatorname{dom}(\boldsymbol{D_A}) = W^{1,2}(\mathbb{R};\mathcal{H}) \cap \operatorname{dom}(\boldsymbol{A_-}).$$

Then with $A_+ = A_- + B_+$, and $A_s = A_- + sB_+$, $s \in [0,1]$, we obtain the main fact, namely, what we call the 'principal trace formula' for the semigroup difference for all $u > 0$,

$$\operatorname{tr}\left(e^{-u\boldsymbol{D_A}\boldsymbol{D_A^*}} - e^{-u\boldsymbol{D_A^*}\boldsymbol{D_A}}\right) = -\left(\frac{u}{\pi}\right)^{1/2}\int_0^1 \operatorname{tr}\left(e^{-uA_s^2}(A_+ - A_-)\right)ds,$$

where, as a result of our assumptions, the operator difference under the trace on the left-hand side is indeed trace-class and the integral on the right-hand side converges.

To appreciate the significance of this principal trace formula, these notes develop several applications. For this we exploit some recent developments in the theory of Krein's spectral shift function from quantum mechanical scattering theory. Going back to the motivating work of Pushnitski, we discuss the proof of a generalisation of his result that relates the spectral shift function of the pair (A_+, A_-) with that for the pair $(\boldsymbol{H}_2, \boldsymbol{H}_1)$ (see Theorem 6.3.1). Then we prove results that relate spectral flow along the path $\{A_- + B(t)\}_{t=-\infty}^{\infty}$ to the Fredholm or Witten index of $\boldsymbol{D_A}$ in the sense that we establish a theorem of Robbin-Salamon type for sufficiently regular paths of self-adjoint Fredholm operators joining operators $A_\pm$ that have some essential spectrum outside zero. An exposition of the theory of the spectral shift function is included.

As the principal trace formula applies also to paths of non-Fredholm operators joining $A_\pm$, we obtain a formula for the Witten index under regularity assumptions on the respective spectral shift functions. This leads us to introduce a notion of 'generalised spectral flow' for such paths and to investigate its properties. Importantly we consider a range of examples to which these more abstract considerations apply. Properties of this extended notion of spectral flow are described, and their proofs exploit the double operator integral technique due originally to [BS66]. We include in these notes some background on double operator integrals including a simplified introduction.

One may think of these results in the following terms. The Atiyah-Singer index theorem gives a topological expression for the Fredholm index of elliptic operators. For the case of Dirac operators on even dimensional compact manifolds, one may formulate the index theorem using a model operator of the same general form as $\boldsymbol{D_A}$. Thus the choice of this 'model operator' is not arbitrary but is adapted to the classical setting. We also note that in supersymmetric quantum mechanics as studied by Witten [Wit82], the Hamiltonians involve the same type of model operator.

For the general operators we consider here, that may have zero in their essential spectrum, there is no topological formula for either the Witten index or generalised spectral flow. Rather, what we find is that scattering theory, which is naturally relevant in understanding the essential spectrum, provides through the spectral shift function, expressions for both of these quantities and also connects them. In order to understand this point of view, one also needs the background material included in these notes on the spectral shift function, the analytic approach to spectral flow, the Witten index and the double operator integral technique.

These lecture notes are based in part on our recent papers [CLPS22] and [CGL+22] and also on lectures given at meetings in BIRS, Santiago and Münster. We have, however, incorporated background from other sources. There are expository sections in each chapter that also include some survey material from the literature going back to the origins of the topics covered. Conversely some technical details are omitted where they may be found in [CLPS22] and do not add substantially to the discussion.

Wollongong, NSW, Australia — Alan Carey
Canberra, ACT, Australia — Galina Levitina

Acknowledgements

A.C. thanks the Erwin Schrödinger International Institute for Mathematical Physics (ESI), Vienna, Austria, for funding support for this collaboration in the form of a Research-in Teams project, 'Scattering Theory and Non-Commutative Analysis', for the duration of June 22 to July 22, 2014. G.L. is indebted to Gerald Teschl and the ESI for a kind invitation to visit the University of Vienna, Austria, for a period of 2 weeks in June/July of 2014. The authors thank BIRS for funding a focussed research group on the topic of these lecture notes in June 2017 and also gratefully acknowledge financial support from the Australian Research Council. A.C. thanks the Humboldt Stiftung and the University of Münster Mathematics Institute for support. G.L. acknowledges the support of Australian government scholarships. We thank Fritz Gesztesy, Jens Kaad, Harald Grosse, Denis Potapov, Fedor Sukochev and Dmitriy Zanin for many conversations on the subject matter of these notes. Finally we are grateful to the referees for carefully reading the manuscript.

Notations

Here we summarise notation that is used without comment in the text.

For a Banach space $\mathcal{X}$, we denote by $\mathcal{L}(\mathcal{X})$ the algebra of all linear bounded operators on $\mathcal{X}$. We denote by 1 the identity mapping on $\mathcal{X}$.

In case when $\mathcal{X} = \mathcal{H}$ is a separable complex Hilbert space $\mathcal{H}$, we use notation $\|\cdot\|$ for the uniform norm.

For an operator $A \in \mathcal{L}(\mathcal{H})$ we denote by $\mathrm{Re}(A)$ (respectively, $\mathrm{Im}(A)$) the real (respectively, imaginary) part of A.

We write $\mathcal{K}(\mathcal{H})$ for the compact operators on $\mathcal{H}$. The Calkin algebra is the quotient algebra $\mathcal{L}(\mathcal{H})/\mathcal{K}(\mathcal{H})$.

The corresponding ℓ_p-based Schatten–von Neumann ideals on $\mathcal{H}$ are denoted by $\mathcal{L}_p(\mathcal{H})$, with associated norm abbreviated by $\|\cdot\|_p$, $p \geq 1$. Moreover, $\mathrm{tr}(A)$ denotes the classical trace of a trace-class operator $A \in \mathcal{L}_1(\mathcal{H})$.

We use the symbols n-lim and s-lim to denote the operator norm limit (i.e., convergence in the topology of $\mathcal{L}(\mathcal{H})$), and the operator strong limit.

If T is a linear operator mapping (a subspace of) a Hilbert space into another, then $\mathrm{dom}(T)$ and $\ker(T)$ denote the domain and kernel (i.e., null space) of T. The closure of a closable operator S is denoted by $\overline{S}$.

The spectrum and resolvent set of a closed linear operator in $\mathcal{H}$ will be denoted by $\sigma(\cdot)$, and $\rho(\cdot)$, respectively. The essential (respectively, point) spectrum of a closed linear operator in $\mathcal{H}$ is denoted by $\sigma_{ess}(\cdot)$ (respectively, by $\sigma_p(\cdot)$).

The spectral projections of a self-adjoint operator S in $\mathcal{H}$ we denote by $E_S(\cdot)$. The space of all (essentially) bounded dE_S-measurable functions is denoted by $L_\infty(\mathbb{R}, dE_S)$.

The notation $[\cdot, \cdot]$ stands for commutator of two operators, that is

$$[A, B] = AB - BA,$$

whenever it is clear how to deal with domains of possible unbounded operators.

For a Fredholm operator T, we denote by $\mathrm{index}(T)$ its Fredholm index.

Unless explicitly stated otherwise, whenever we write $L_p(\mathbb{R}^d)$ ($L_p(0, \infty)$ etc.) we assume the classical Lebesgue measure on $\mathbb{R}^d$ ($(0, \infty)$ etc.). The space of all

Schwartz function on $\mathbb{R}^d$ is denoted by $S(\mathbb{R}^d)$ and the Sobolev spaces are denoted by $W^{p,q}(\mathbb{R}^d)$. The space of all Lipschitz functions on $\mathbb{R}$ is denoted by $Lip(\mathbb{R})$.

The space of all functions with continuous derivative up to order n, $n \in \mathbb{N} \cup \{\infty\}$ is denoted by $C^n(\mathbb{R})$ (or $C^n(a,b)$, $a < b$). The space of all compactly supported functions derivative up to order n, $n \in \mathbb{N} \cup \{\infty\}$ is denoted by $C_0^n(\mathbb{R})$ (or $C^n(a,b)$, $a < b$). If all derivatives up to order n are also bounded functions, then the space is denoted by $C_b^n(\mathbb{R})$ ($C_b^n(a,b)$, respectively).

By $L_2(\mathbb{R}, \mathcal{H})$, we denote the Hilbert space of all $\mathcal{H}$-valued Bochner square integrable functions on $\mathbb{R}$. The space $W^{p,q}(\mathbb{R}^d, \mathcal{H})$ denotes the Sobolev spaces of $\mathcal{H}$-valued functions. We will make use of that bold face operators $\boldsymbol{T}$ act in the Hilbert space $L_2(\mathbb{R}; \mathcal{H})$ and typically represent operators associated with a family of operators $\{T(t)\}_{t\in\mathbb{R}}$ in $\mathcal{H}$, defined by

$$\begin{aligned}
&(\boldsymbol{T} f)(t) = T(t) f(t) \text{ for a.e. } t \in \mathbb{R},\\
&f \in \operatorname{dom}(\boldsymbol{T}) = \Big\{ g \in L_2(\mathbb{R}; \mathcal{H}) \,\big|\, g(t) \in \operatorname{dom}(T(t)) \text{ for a.e. } t \in \mathbb{R};\\
&\qquad t \mapsto T(t)g(t) \text{ is (weakly) measurable; } \int_{\mathbb{R}} \|T(t)g(t)\|_{\mathcal{H}}^2 dt < \infty \Big\}.
\end{aligned}$$

We denote by $\lfloor\cdot\rfloor$ the floor function on $\mathbb{R}$ (i.e. $\lfloor x\rfloor$ is the largest integer which is less than or equal $x \in \mathbb{R}$) and for a real number $x \in \mathbb{R}$ the notation $\{x\}$ stands for the fractional part of x given by $x - \lfloor x\rfloor$.

For a function $f : \mathbb{R} \to \mathbb{R}$, we denote by $f(0_+)$ and $f(0_-)$ the left- and right-hand limit of f at 0, if existent. The left (respectively, right) value of f at zero in the Lebesgue sense (see Definition 6.4.8) is denoted by $f_L(0-)$ ($f_L(0+)$, respectively).

The characteristic function of a Borel set A is denoted by χ_A. We employ the abbreviations

$$g_z(x) = x(x^2 - z)^{-1/2}, \;\; z \in \mathbb{C}\backslash[0,\infty), \quad g(x) = g_{-1}(x), \quad x \in \mathbb{R}.$$

We denote by erf the error function

$$\operatorname{erf}(x) = \frac{2}{\pi^{1/2}} \int_0^x e^{-y^2} dy, \quad x \in \mathbb{R}.$$

Contents

Chapter 1
Introduction

1.1 Motivation and Background

In writing these notes we began with the notion of giving a detailed exposition of the broad area captured by the phrase 'index theory beyond the Fredholm case'. However this would be a very ambitious project. To keep this to a manageable text we decided to focus here on one aspect of this topic which is nevertheless broad enough to touch on all previous work.

The history of this line of research on 'index theory' for non-Fredholm operators commenced some time ago in a paper by R. W. Carey and J. Pincus [CP86] that focussed on bounded operators and then, for the unbounded case, independently by F. Gesztesy and B. Simon in [GS88]. In both of these papers this extended notion of index (intended to go beyond the Fredholm case) is expressed in terms of Krein's spectral shift function from scattering theory. In the former paper the starting point is an operator T on a separable Hilbert space $\mathcal{H}$ with the property that the commutator $TT^* - T^*T$ is trace-class. In the latter paper the problem is stated in terms of a generalised McKean-Singer formula that applies to some unbounded operators motivated by examples in [BGG⁺87]. The passage from the unbounded picture to the bounded one is straightforward and is explained below.

For many years there was little further development of this functional analytic formulation of a generalised index for non-Fredholm operators. One obstacle was that it was not appreciated in these early days that the assumptions made in these early papers ruled out applications to Hamiltonians in higher space dimensions.

Starting from a different viewpoint a new attempt to investigate higher dimensional examples was initiated in [CGK15]. In that paper the existence of a class of non-trivial examples (Dirac type operators) was found by using a new condition that replaced the Carey-Pincus trace-class commutator assumption. This new condition involved higher Schatten classes. As we explain later the general framework described here applies to [CGK15] and was in fact partly motivated by that paper. The discussion of [CGK15] illustrates in the unbounded context the appropriate

A. Carey, G. Levitina, *Index Theory Beyond the Fredholm Case*, Lecture Notes in Mathematics 2323, https://doi.org/10.1007/978-3-031-19436-8_1

generalisation of the Carey-Pincus framework to the case where one replaces the trace ideal by other Schatten ideals. In [CK17] it was shown how to generalise the work of Carey-Pincus using cyclic homology. We will only sketch this direction in these notes. The important observation to be drawn from that work is that the trace-class commutator condition is relevant only for low dimensional manifolds but does not apply in higher dimensions.

The index studied in [CP86], and [GS88] is not invariant under compact perturbations (respectively, relatively compact perturbations) and hence has no relationship to topology or to K-theory. There has been much effort expended in order to understand, in a broader more applicable framework, these early papers. We will touch on that work, where relevant, in various places in these notes.

The starting point for us was the paper [GLM+11]. It was there that a connection was made between the Fredholm index and the spectral flow when the operators in question had essential spectrum.

Now the usual approach to index theory of elliptic (Fredholm) operators with essential spectrum, as occurs with non-compact manifolds, would be to replace the original index problem by one where the operators had discrete spectrum only. However if the operators being studied are Hamiltonians of quantum systems then one is usually interested in more detailed spectral information than just the index (for example the study of resonances). Tinkering with the Hamiltonian so as to eliminate the essential spectrum will destroy important information.

The idea of working directly with operators having essential spectrum was the innovation introduced in [GLM+11] building on the original paper [Pus08] of Pushnitski. The main result of [GLM+11] was to produce a formula relating the Fredholm index of a certain model operator with essential spectrum to a spectral flow problem. This introduction of the spectral flow into the picture opened a new chapter that we were motivated to study and which led to the present set of notes.

The notion of spectral flow along a path of self-adjoint elliptic operators was introduced in Atiyah-Patodi-Singer [APS76]. There they explained how one might compute it by using a 'suspension' trick that produces a first order elliptic operator on a manifold of one higher dimension whose Fredholm index was equal to the spectral flow along the original path. Analogous theorems were later studied by other methods (see for example [BBW93]) culminating in a definitive treatment for certain self-adjoint differential operators with compact resolvent in a paper of Robbin-Salamon, [RS95]. Their paper has implications for many applications including those to Morse theory, Floer homology, Morse and Maslov indices, Cauchy-Riemann operators, and all the way to oscillation theory.

Spectral flow in the early works [APS76, RS95] is treated in a topological fashion and in particular, K-theoretically in [RS95]. A new, purely analytic approach to spectral flow was introduced in [Phi96]. Combined with the viewpoint of [Get93] this produced analytic formulas for the spectral flow [CP98, CP04], providing an analytic counterpoint to the previous K-theoretic and intersection number approaches. This opened the way to an analytic formulation of the Robbin-Salamon theorem. A first step in this direction was the introduction in [BCP+06] of an operator theoretic trace formula that gives a version of the Robbin-Salamon result proved in the setting of Atiyah's L^2 index theorem (that is it uses the notion a 'von Neumann spectral flow' that arises in semi-finite von Neumann algebras). All of these results however assume that the path of unbounded Fredholm operators along which the spectral flow is computed have compact resolvents (in the sense of semifinite von Neumann algebras).

With the exception of [BCP⁺06], the focus in mathematics on these questions has been mainly on geometrically defined operators associated to compact manifolds. Physicists, however, are interested in the case of non-compact manifolds and non-geometric examples. The stumbling block for index formulas in that situation is the presence of essential spectrum. Motivated by ideas from scattering theory an approach to a Robbin-Salamon type result for paths of self-adjoint Fredholm operators with some essential spectrum was initiated in [Pus08], followed by [GLM⁺11]. However, the key assumption in [GLM⁺11] is that the spectral flow is only considered along paths between self-adjoint operators that differ by a so-called relatively trace-class perturbation. This latter assumption is satisfied by certain pseudo-differential perturbations of a fixed differential operator but does not apply to paths of differential operators even in one dimension. As a result, the promising start made in [GLM⁺11] to generalising the Robbin-Salamon theorem (so that it applied to operators with some essential spectrum as occurs in the non-compact manifold case), ran into difficulties.

Partial results (applicable to differential operators only in lower dimensions) have been obtained in the intervening years [CGL⁺16b, CGLS16a, CGG⁺16]. A significant advance that motivates these notes was made in the article [CGK16]. There, a trace formula is proved, again motivated by Benameur et al. [BCP⁺06] and Gesztesy et al. [GLM⁺11] (but by completely different methods than we employ here), that specialises under compact resolvent assumptions to give another version of the Robbin-Salamon theorem. More importantly however the formula applies even to situations where the operators considered are not Fredholm. It applies whether or not the operators in the path have zero in the essential spectrum. However, what is important is that in [CGK16] the trace formula was related to the spectral flow only in the setting of unitarily equivalent endpoints with purely discrete spectra. We explain how to remove this restriction in these notes.

Relevant to the issue of allowing essential spectrum is that some time ago Witten [Wit82] gave a proposal for extending index theory beyond the Fredholm setting. For certain Dirac type operators (with essential spectrum) Witten's ideas could be shown to produce index theorems [BGG⁺87, GS88]. Some of this early work also involves non-Fredholm operators and it was revisited in [CGP⁺17] and related to Pushnitski's work under the same relative trace-class assumption as in [GLM⁺11].

At that time the question of how to handle Dirac operators on higher dimensional spaces remained unresolved. In [CGLS16a] we began to complete the program started in [GLM⁺11]. The new ingredient in [CGLS16a] is an approximation technique that enabled the main theorems in [GLM⁺11] to be extended to more general examples where relatively trace-class hypotheses were violated. It is this approximation technique that underpins the advances described in these notes.

Finally, further motivation for presenting an extended account of these developments stems from attempts to use the notion of spectral flow in condensed matter theory, for example, [Sto96]. In this application Dirac type operators are used as Hamiltonians and the spectral flow along paths of such operators provides information of physical interest. In some cases the operators in question are Fredholm with essential spectrum. Thus the Robbin-Salamon theorem does not apply. In other instances the paths of operators considered violate an important assumption in [RS95] namely, that the endpoints of the path of operators along which the spectral flow is to be calculated are invertible. In these notes we are able to provide information on the spectral flow in this situation. We also mention that, in some examples using physical models, the analytic spectral flow definition used in our discussion does not apply (as the operators in the path are not necessarily Fredholm). Our approach enables us to prove results in this non-Fredholm setting (see Theorem 6.4.12).

We summarise our point of view by remarking that whereas the Atiyah-Singer theorem [APS76] relates the analytical index to the spectral flow via the topological index our results express both the analytic index and the spectral flow in terms of quantities from quantum mechanical scattering theory. We note further that in the non-Fredholm situation that we study here we cannot necessarily expect topological formulas. This may be understood from the fact that in this non-Fredholm situation we investigate properties that hold modulo the trace-class whereas the topological results follow because one works modulo compact operators. It is surprising nevertheless that we are able to replace the Fredholm index by the Witten index and obtain a generalisation of spectral flow in our framework.

In these notes we describe a number of major recent advances. First, we formulate our discussion in a purely operator theoretic fashion and hence we may apply our approach in non-geometrical settings. Second in the setting of the Robbin-Salamon theorem we show that when one drops the assumption of invertible endpoints for the path one still obtains an index formula except that the Witten index replaces the Fredholm index. Third, we allow essential spectrum in the case of paths of self-adjoint Fredholm operators. Fourth, we show that we obtain information even in the case where the path of self-adjoint operators is not Fredholm. It is in this final situation that substantial results from quantum mechanical scattering theory are essential. And finally our results lead us to introduce a notion of a generalised spectral flow that provides a counterpart to the Witten index in the odd dimensional case.

1.2 An Overview of Recent Results

We now briefly discuss the setting for the recent results explained in detail in these notes. We defer the discussions of methods and applications to the following section. For the purpose of an accessible introduction we consider here a simplified situation of the general setting under which we prove the results.[1]

[1] Notation not explained in the text may be found in the Notations.

We start with a self-adjoint unbounded operator A_- densely defined on a separable complex Hilbert space $\mathcal{H}$ and suppose that B is a self-adjoint bounded perturbation of A_-. If the perturbation B is a relative trace-class perturbation of A_-, that is, $B(A_- + i)^{-1}$ is a trace-class operator, then the main assumption in [GLM+11, CGP+17] is satisfied. Hence, the results summarised in the previous subsection are known to be true. However, the critical fact for partial differential operators in general is that perturbations by lower order operators satisfy relative Schatten–von Neumann class constraints but not relative trace-class constraints. To describe this, suppose for example that we consider A_- to be the flat space Dirac operator acting in $L_2(\mathbb{R}^d) \otimes \mathbb{C}^{n(d)}$ and the perturbation B to be given by the multiplication operator by a smooth, $n(d) \times n(d)$ matrix-valued bounded function

$$F : \mathbb{R}^d \to M^{n(d)\times m(d)}(L_\infty(\mathbb{R}) \cap C^\infty(\mathbb{R})).$$

Under suitable decay conditions at infinity for F, the product $B(1 + A_-^2)^{-s/2}$ is trace-class for $s > d$ and for no smaller value of s (see [Sim05, Remark 4.3]). Therefore, the case where we may set $s = 1$ and obtain a relative trace-class perturbation requires we work in dimension zero. The only way that we know of to obtain a relative trace-class perturbation in d-dimensions is to replace F by certain pseudo-differential operators. It eventuates, as we explain in later sections, that an appropriate choice of pseudo-differential perturbations can be used to approximate the kind of perturbation of Dirac operators that arise in examples. In this way the results of [GLM+11] and [CGP+17] can be brought to bear on the higher dimensional situation.

Thus, if we want to include multidimensional examples in our setting the assumption on the perturbation B should be that the operator $B(A_- + i)^{-p-1}$ is a trace-class operator for some fixed $p \in \mathbb{N} \cup \{0\}$. This assumption includes, in particular, the assumptions of [Pus08] and [GLM+11, CGP+17]. However, what is more important, in contrast to these three papers, our assumption is satisfied by Dirac type operators on certain non-compact manifolds (see [CGRS14]). See Sect. 7.1 below for the example of the Dirac operator on $\mathbb{R}^d$, $d \in \mathbb{N}$.

To discuss the connection to index theory, we introduce now the 'suspension' operator $\boldsymbol{D}_A$ as in [APS76, RS95]. Let $\theta : \mathbb{R} \to \mathbb{R}$ be a smooth function (with integrable positive derivative) interpolating between zero and one in the sense that $\lim_{t\to-\infty} \theta(t) = 0$ and $\lim_{t\to\infty} \theta(t) = 1$. Denoting by M_θ the operator on $L_2(\mathbb{R})$ of multiplication by θ, we introduce the operator $\boldsymbol{D}_A$ in the Hilbert space $L_2(\mathbb{R}) \otimes \mathcal{H}$ by

$$\boldsymbol{D}_A = \frac{d}{dt} \otimes 1 + 1 \otimes A_- + M_\theta \otimes B. \tag{1.1}$$

Here the operator d/dt in $L_2(\mathbb{R})$ is the differentiation operator with domain being the Sobolev space $W^{1,2}(\mathbb{R})$, so that $\boldsymbol{D}_A$ is defined on $W^{1,2}(\mathbb{R}) \otimes \mathrm{dom}(A_-)$. For ease of notation we will usually identify $L_2(\mathbb{R}) \otimes \mathcal{H}$ and $L_2(\mathbb{R}; \mathcal{H})$.

In order to relate the index theory of the operator $\boldsymbol{D_A}$ with the spectral flow (when defined) of the family $\{A_- + \theta(t)B\}_{t\in\mathbb{R}}$ we establish firstly the first primary result of this current notes, the *principal trace formula*. Namely, under some additional mild assumptions for any $t > 0$ we prove the following relation (see Theorem 6.2.2 below):

$$\operatorname{tr}\left(e^{-t\boldsymbol{D_A}\boldsymbol{D_A^*}} - e^{-t\boldsymbol{D_A^*}\boldsymbol{D_A}}\right) = -\left(\frac{t}{\pi}\right)^{1/2} \int_0^1 \operatorname{tr}\left(e^{-tA_s^2}B\right)ds, \tag{1.2}$$

$$A_s = A_- + sB, \quad s \in [0,1],$$

noting that our hypotheses guarantee both sides of the relation are well-defined. Here, tr denotes the classical trace on the algebra $\mathcal{L}(\mathcal{H})$ of all bounded linear operator on a Hilbert space $\mathcal{H}$.

The key point to note is that this is an operator identity that makes no assumptions on the spectrum of the operators on either side of the relation. One of our key objectives is to develop the consequences of this fact, principally by using scattering theory methods. We discuss these methods in the next section.

We note that if we impose the assumption that the operators A_- and $A_+ := A_- + B$ have discrete spectrum and are unitarily equivalent and invertible then the operators $A_\pm$ and $\boldsymbol{D_A}$ are Fredholm and in that case the left-hand side of (1.2) is the Fredholm index of $\boldsymbol{D_A}$ [GS88] while the right hand side is the spectral flow along the path $\{A_- + \theta(t)B\}_{t\in\mathbb{R}}$ [ACS07]. Thus the principal trace formula entails a version of the Robbin-Salamon theorem.

However, if we assume that the endpoints $A_\pm$ are not invertible, the principal trace formula remains true, but the left-hand side is no longer the Fredholm index since the operator $\boldsymbol{D_A}$ is no longer Fredholm. However the right-hand side will still be the spectral flow if $A_\pm$ have discrete spectrum. As the right-hand side is independent of t in this case [CP04] so too is the left-hand side and as discussed below, it is the Witten index of $\boldsymbol{D_A}$.

If the operator A_- has some essential spectrum then neither the left hand side nor the right hand side of the principal trace formula can be proved to be independent of t. In fact this problem is already apparent in the older work of Callias [Cal78]. There are then two possible asymptotic quantities that we might consider. The first, the limit as $t \to 0$ on the left-hand side of the principal trace formula, gives what has been referred to in the physics literature as the anomaly. In the context of the Atiyah-Singer index theorem it gives the local form of the theorem involving integrals of characteristic classes. There is evidence [CGK15] that for Dirac type operators with some essential spectrum the anomaly can be expressed in terms of a local formula of the same type as arises in the Atiyah-Singer index theorem.

Our main interest in these notes lies in the second limit as $t \to \infty$ of (1.2). On the left hand side this limit, if it exists, has been termed the Witten index [GS88]. In the case of the right-hand side this limit has not been previously investigated. We provide here arguments that allow us to regard this limit of the right-hand side as a generalisation of spectral flow to the non-Fredholm setting. In particular, if the endpoints $A_\pm$ are Fredholm operators, this generalised spectral flow gives the classical spectral flow. However, we will show in the final section using a simple example that the anomaly, spectral flow and the Witten index do not coincide in general.

The study of the limit $t \to \infty$ in the principal trace formula is made possible by exploiting Krein's spectral shift function from quantum mechanical scattering theory. That is, both limits are computed by the respective spectral shift functions. We include in these notes an outline of the history of the spectral shift function and some details of recent results that probe its properties. From results on the spectral shift function a number of applications of the principal trace formula emerge. We include here a generalisation of the original formulas of Pushnitski relating the spectral shift function for the pair (A_+, A_-) with that for the pair $(\boldsymbol{D}_{\boldsymbol{A}}^* \boldsymbol{D}_{\boldsymbol{A}}, \boldsymbol{D}_{\boldsymbol{A}} \boldsymbol{D}_{\boldsymbol{A}}^*)$. We also explain consequential results for the Witten index, and for the spectral flow and its generalisations.

In the body of the text we will formulate results for model operators $\boldsymbol{D}_A$ that are more general than the one discussed above. The form of these model operators is dictated on the mathematical side by the Atiyah-Singer index theorem for Dirac type operators on even dimensional compact manifolds and on the physics side by Hamiltonians in supersymmetric quantum mechanics as introduced by Witten [Wit82]. In the next section we will elaborate on this point.

1.3 Discussion of the Methods and the Applications in These Notes

The principal trace formula (1.2) is an operator theoretic identity that makes no Fredholm type assumptions on the operators in question. For this reason it may be used to obtain information when the model operator $\boldsymbol{D}_A$ is non-Fredholm. We will make use of the ideas that were previously introduced in [CGP+17]. In that paper the model operator $\boldsymbol{D}_{\boldsymbol{A}}$ in $L_2(\mathbb{R}; \mathcal{H})$ was studied without Fredholm assumptions.

In this setting we replace the Fredholm index by the so-called Witten index. Recall [GS88] that the semigroup (heat kernel) regularized Witten index, denoted $W_s(\boldsymbol{D}_{\boldsymbol{A}})$, of the operator $\boldsymbol{D}_{\boldsymbol{A}}$ is defined by

$$W_s(\boldsymbol{D}_{\boldsymbol{A}}) = \lim_{t\uparrow\infty} \operatorname{tr}\left(e^{-t\boldsymbol{D}_{\boldsymbol{A}}^*\boldsymbol{D}_{\boldsymbol{A}}} - e^{-t\boldsymbol{D}_{\boldsymbol{A}}\boldsymbol{D}_{\boldsymbol{A}}^*}\right),$$

whenever the limit exists.

In the particular case, when the operator $\boldsymbol{D}_{\boldsymbol{A}}$ is Fredholm, one has consistency with the Fredholm index, index$(\boldsymbol{D}_{\boldsymbol{A}})$ of $\boldsymbol{D}_{\boldsymbol{A}}$ [GS88], that is,

$$\operatorname{index}(\boldsymbol{D}_{\boldsymbol{A}}) = W_s(\boldsymbol{D}_{\boldsymbol{A}}).$$

There is also a resolvent regularised Witten index described in [BGG⁺87, CGP⁺17] and references therein, introduced via the difference of resolvents of $\boldsymbol{D}_A^*\boldsymbol{D}_A$ and $\boldsymbol{D}_A\boldsymbol{D}_A^*$. However, in the setting of differential operators on higher dimensional manifolds this difference is typically not a trace-class operator. Hence, following [CGK15], we introduced k-th resolvent regularised Witten index $W_{k,r}(\boldsymbol{D}_A)$ of $\boldsymbol{D}_A$ by setting

$$W_{k,r}(\boldsymbol{D}_A) = \lim_{\lambda\uparrow 0}(-\lambda)^k \operatorname{tr}\left((\boldsymbol{D}_A^*\boldsymbol{D}_A - \lambda)^{-k} - (\boldsymbol{D}_A\boldsymbol{D}_A^* - \lambda)^{-k}\right)$$

whenever this limit exists. In the terminology of [CGK15] this is the limit as $\lambda \uparrow 0$ of the homological index of $\boldsymbol{D}_A$.

We now explain the connection with the Carey-Pincus work [CP86]. They start with a bounded operator T on a Hilbert space $\mathcal{H}$ with $TT^* - T^*T$ trace-class. The appropriate generalisation of this condition to higher Schatten classes is to suppose that T satisfies:

$$(1 - T^*T)^n - (1 - TT^*)^n \tag{1.3}$$

is in the trace-class. For $n = 1$ this condition reduces to the trace-class commutator condition.

Unbounded operators can sometimes be mapped to a pair (T, T^*) satisfying a Schatten class condition as above. A source of examples is studied in [CGK15]. These arise from taking a Dirac operator acting on L^2-sections of the spin bundle over $\mathbb{R}^{2n}$ and coupling it to a connection. If we write this operator as $\mathcal{D}$ then we pass to the bounded picture using the map

$$\mathcal{D} \to \mathcal{D}(1 + \mathcal{D}^2)^{-1/2}.$$

There is a natural grading on the Hilbert space $\mathcal{H}$ of L^2 sections of the spin bundle tensored with an auxiliary bundle equipped with the connection. This grading is an operator on $\mathcal{H}$ which is self-adjoint, bounded and satisfies $\gamma^2 = 1$. The spectral representation of γ allows us to split the Hilbert space into two subspaces because we can write $\gamma = \begin{pmatrix} I & 0 \\ 0 & -I \end{pmatrix}$.

The Dirac operator $\mathcal{D}$ anticommutes with γ and so can be represented, in the spectral representation of γ as $\mathcal{D} = \begin{pmatrix} 0 & \mathcal{D}_- \\ \mathcal{D}_+ & 0 \end{pmatrix}$, where $\mathcal{D}_- = \mathcal{D}_+^*$. In this notation $\mathcal{D}_+$ plays the role of the model operator $\boldsymbol{D}_A$ introduced previously.

For z in the intersection of the resolvent sets one may show that

$$(z + \mathcal{D}_-\mathcal{D}_+)^{-n} - (z + \mathcal{D}_+\mathcal{D}_-)^{-n}$$

is in the trace-class for sufficiently large n. We explain the argument in Chap. 6.

To reformulate this situation in terms of bounded operators we introduce

$$T = \mathcal{D}_+(1 + \mathcal{D}_-\mathcal{D}_+)^{-1/2}.$$

By scaling $\mathcal{D}$, that is, replacing it by $\lambda^{-1/2}\mathcal{D}$ we obtain a one parameter family of pairs T_λ, T_λ^*.

The scaling limit as $\lambda \to \infty$ of

$$\operatorname{tr}((1 - T_\lambda^* T_\lambda)^n - (1 - T_\lambda T_\lambda^*)^n) \tag{1.4}$$

is referred to as the 'anomaly' in the mathematical physics literature. On the other hand the scaling limit as $\lambda \to 0$ that is studied in [GS88] is exactly the resolvent form of the Witten index.[2]

The expression (1.4) is easily seen to be just the homological index of [CGK16]. Later in these notes we will sketch the homological interpretation of this higher order Carey-Pincus theory following [CK17]. It is a basic fact that for Dirac type operators on compact manifolds the expression (1.4) is λ independent being the Fredholm index of T. However examples show that when T has essential spectrum one expects that (1.4) is λ dependent as may be seen from the old work of Callias [Cal78]. Readers familiar with the Calderón formula for the Fredholm index of an operator T will recognise (1.4).

In general (i.e., if $\boldsymbol{D}_{\boldsymbol{A}}$ is not Fredholm), $W_s(\boldsymbol{D}_{\boldsymbol{A}})$ is not necessarily integer-valued; in fact, one can show that it can take on any prescribed real number. The intrinsic value of $W_s(\boldsymbol{D}_{\boldsymbol{A}})$ then lies in its stability properties with respect to additive perturbations, analogous to stability properties of the Fredholm index. Indeed, as long as one replaces the familiar relative compactness assumption on the perturbation in connection with the Fredholm index, by appropriate relative trace-class conditions in connection with the Witten index, stability of the Witten index was proved in [BGG+87] and [GS88].

We will not go over the extensive history associated with the Witten index but refer the reader to [CGP+17] for a full bibliography. We remark however that the geometric significance of the Witten index when it does not coincide with anomaly is not clear.

As can be seen from the definition of the semigroup regularised Witten index $W_s(\boldsymbol{D}_{\boldsymbol{A}})$ and the principal trace formula (1.2), we can compute $W_s(\boldsymbol{D}_{\boldsymbol{A}})$ by taking the limit of the right-hand side of principal trace formula as $t \to \infty$. To describe this limit we use methods of scattering theory, namely the spectral shift function of M.G. Krein. For a full exposition of the theory of Krein's spectral shift function we refer

[2] It was Gesztesy-Simon who discovered the connection between Witten's ideas and the spectral shift function and hence the link to Carey-Pincus. There is an extensive literature on the Witten index and supersymmetric quantum mechanics some of which is touched on here. We note that the existence of the grading operator γ is what is termed 'supersymmetry' in this quantum mechanical setting.

to [Yaf92, Chapter 8] while for those aspects relevant to this article we include an exposition later in these notes. One may also consult our review article [CGLS16b].

The spectral shift function was introduced by M.G. Krein in [Kre53], who was inspired by results of Lifshitz on Hamiltonians of a lattice model in quantum mechanics. Krein proved that for two self-adjoint operators H and H_0 with common domain and the perturbation $H - H_0$ of trace-class, there exists a unique (Lebesgue) integrable function $\xi(\cdot; H, H_0)$ on $\mathbb{R}$ satisfying the so-called Lifshitz-Krein trace formula

$$\operatorname{tr}\big(f(H) - f(H_0)\big) = \int_{\mathbb{R}} f'(\lambda)\xi(\lambda; H_0, H_0 + V)d\lambda,$$

for a large class of functions f.

Krein's work can be applied (with some further definitions that we omit for the purposes of this introductory discussion) to the operators introduced above, that is, to the pairs of operators (A_+, A_-) and $(\boldsymbol{D}_{\boldsymbol{A}}^*\boldsymbol{D}_{\boldsymbol{A}}, \boldsymbol{D}_{\boldsymbol{A}}\boldsymbol{D}_{\boldsymbol{A}}^*)$. Hence, using the Lifshitz-Krein trace formula for the left-hand side of principal trace formula (1.2) we can write

$$\operatorname{tr}(e^{-t\boldsymbol{D}_{\boldsymbol{A}}^*\boldsymbol{D}_{\boldsymbol{A}}} - e^{-t\boldsymbol{D}_{\boldsymbol{A}}\boldsymbol{D}_{\boldsymbol{A}}^*}) = -t\int_0^\infty \xi(\lambda; \boldsymbol{D}_{\boldsymbol{A}}^*\boldsymbol{D}_{\boldsymbol{A}}, \boldsymbol{D}_{\boldsymbol{A}}\boldsymbol{D}_{\boldsymbol{A}}^*)e^{-t\lambda}\, d\lambda.$$

That is, the left-hand side of the principal trace formula can be rewritten as the Laplace transform of the spectral shift function $\xi(\cdot; \boldsymbol{D}_{\boldsymbol{A}}^*\boldsymbol{D}_{\boldsymbol{A}}, \boldsymbol{D}_{\boldsymbol{A}}\boldsymbol{D}_{\boldsymbol{A}}^*)$.

For the right-hand side of the principal trace formula, the Lifshitz-Krein trace formula can not be used directly. However, exploiting results from [CGL+16a], we prove the equality

$$\int_0^1 \operatorname{tr}\Big(e^{-tA_s^2}(A_+ - A_-)\Big)ds = \int_{\mathbb{R}} \xi(s, A_+, A_-)e^{-ts^2}ds, \quad t > 1,$$

and therefore, as a corollary of the principal trace formula, the uniqueness theorem for the Laplace transform and its simple properties we conclude that the spectral shift functions $\xi(\cdot; \boldsymbol{D}_{\boldsymbol{A}}^*\boldsymbol{D}_{\boldsymbol{A}}, \boldsymbol{D}_{\boldsymbol{A}}\boldsymbol{D}_{\boldsymbol{A}}^*)$ and $\xi(\cdot; A_+, A_-)$ are related via a so-called Pushnitski formula (see Theorem 6.3.1 below)

$$\xi(\lambda; \boldsymbol{D}_{\boldsymbol{A}}^*\boldsymbol{D}_{\boldsymbol{A}}, \boldsymbol{D}_{\boldsymbol{A}}\boldsymbol{D}_{\boldsymbol{A}}^*) = \frac{1}{\pi}\int_{-\lambda^{1/2}}^{\lambda^{1/2}} \frac{\xi(\nu; A_+, A_-)\, d\nu}{(\lambda - \nu^2)^{1/2}}, \qquad \text{a.e. } \lambda > 0,$$

which generalises [Pus08] and [GLM+11].

Pushnitski's formula together with some Tauberian theorems allows us to establish one of the main applications in these notes. This application relates the Witten index of the operator $\boldsymbol{D}_{\boldsymbol{A}}$ with the value at zero (in a Lebesgue sense) of the spectral shift function $\xi(\cdot; A_+, A_-)$ (see Sect. 6.4 for the precise definition). We use the notation $\xi_L(0_+; A_+, A_-)$ and $\xi_L(0_-; A_+, A_-)$, respectively, to denote the left and right hand value of $\xi(\cdot; A_+, A_-)$ at zero in the sense of Lebesgue.

Theorem 1.3.1 *Assume Hypothesis 3.2.5. Assume that 0 is a right and a left Lebesgue point of $\xi(\cdot\,; A_+, A_-)$. Then the semigroup regularised Witten index $W_s(\boldsymbol{D}_{\boldsymbol{A}})$ as well as the k-th resolvent regularised index $W_{k,r}(\boldsymbol{D}_{\boldsymbol{A}})$ exist for sufficiently large $k \in \mathbb{N}$ and equal*

$$W_s(\boldsymbol{D}_{\boldsymbol{A}}) = W_{k,r}(\boldsymbol{D}_{\boldsymbol{A}}) = [\xi_L(0_+; A_+, A_-) + \xi_L(0_-; A_+, A_-)]/2. \tag{1.5}$$

As one-dimensional examples show [CGL+16b, CGG+16], the equality in Theorem 1.3.1 holds even in the case when the endpoints $A_\pm$ have purely absolutely continuous spectra coinciding with the whole real line. Our investigation of multi-dimensional examples demonstrates that, in many cases, the spectral shift function $\xi(\cdot; A_+, A_-)$ is sufficiently regular at zero for the abstract analysis described in these notes to apply, and therefore, equality (1.5) holds.

As shown in [ACS07] for operators $A_\pm$ with compact resolvent (in the sense of general semifinite von Neumann algebras), the spectral shift function $\xi(\cdot; A_+, A_-)$ at zero is equal to the spectral flow up to kernel correction terms. However, for our primary example of Dirac operators on $\mathbb{R}^d$, the operators $A_\pm$ do not have compact resolvent. Thus, in our setting, [ACS07] is not applicable.

Nevertheless, in the case when the endpoints $A_\pm$ are Fredholm and for a suitable path $\{A(t)\}_{t\in\mathbb{R}}$ joining A_+ and A_-, we prove that the spectral flow for $\{A(t)\}_{t\in\mathbb{R}}$, defined using the analytic approach of [Phi96], is again equal to the value of the spectral shift function $\xi(\cdot; A_+, A_-)$ at zero modulo the correction terms. In particular, employing Theorem 1.3.1, we obtain the following generalisation of the Robbin-Salamon formula.

Theorem 1.3.2 *Assume Hypothesis 3.2.5 and assume, in addition, that the endpoint $A_\pm$ are Fredholm operators. Then the spectral flow $\mathrm{sf}(\{A(t)\}_{t\in\mathbb{R}})$ along the path $\{A(t)\}_{t\in\mathbb{R}}$ exists, the Witten index $W_s(\boldsymbol{D}_{\boldsymbol{A}})$ of the operator $\boldsymbol{D}_{\boldsymbol{A}}$ exists and we have the equality*

$$W_s(\boldsymbol{D}_{\boldsymbol{A}}) = \mathrm{sf}(\{A(t)\}_{t\in\mathbb{R}}) - \frac{1}{2}[\dim(\ker(A_+)) - \dim(\ker(A_-))].$$

In the particular case, when the path $\{A(t)\}_{t\in\mathbb{R}}$ consists of operators with purely discrete spectra and the endpoints are unitarily equivalent we recover a version of the result of [RS95]. Interestingly though, our formula generalises also results of [Pus08] and [GLM+11]. We also note [CGP+15], in which it is shown that for unitarily equivalent endpoints, where the unitary satisfies additional constraints, one may obtain a residue formula for the spectral flow between $A_\pm$. The limiting process used there is analogous in some ways to the limiting process we use here in that we take the parameter t in the principal trace formula to infinity.

This extension of previous results to the case where A_- is Fredholm, but there are no restrictions on the kernels of either A_- or A_+, is likely to have applications in condensed matter theory. We have in mind models of, for example, topological phases of matter, where $A_{\pm}$ are Hamiltonians and the case where they have kernels is precisely the most interesting one.

Now we turn to the case where $A_{\pm}$ need not be Fredholm. This puts us outside existing theory except for [CGP+17] and [CGLS16b] where relatively trace-class perturbation assumptions are made. Given that almost nothing is known about this situation these notes go into some detail on what can be said at this time. The most important of these is the principal trace formula (1.2) itself.

In the non-Fredholm setting, the left-hand side of the principal trace formula leads to the Witten index. Hence, a natural question is to ask what the right-hand side of the principal trace formula represents in the non-Fredholm case. In the absence of compact resolvent assumptions our ideas are inspired by an old example of one of us (ALC) and Harald Grosse that may be found in an unpublished manuscript having its roots in investigations of gauge transformations in fermionic quantum field theories in the 1980s [CHO82]. We described the example in [CGG+16]. This example suggested to us that the limit of the right-hand side of the principal trace formula as $t \to \infty$, when it exists, is a generalisation of the notion of spectral flow to the non-Fredholm situation just as the Witten index generalises the Fredholm index.

However, to show that this limit of the right-hand side of the principal trace formula can be regarded as a generalised spectral flow, we have to establish some properties of the integral

$$\int_0^1 \operatorname{tr}\left(e^{-tA_s^2}B\right)ds, \quad A_s = A_- + sB, \quad t > 0 \tag{1.6}$$

on the right-hand side of the principal trace formula. For the path of Fredholm operators (with a certain smoothness condition), it is known that the integral in (1.6) represents the integral formula for the spectral flow [CP98]. As spectral flow is a homotopy invariant of the space of Fredholm operators, we anticipate by analogy, to obtain here a weakened form of homotopy invariance for the integral (1.6). We explain in these notes that in general this integral is independent of the choice of sufficiently smooth path joining $A_{\pm}$ as long as this path remains in an affine space of 'admissible' perturbations (we make this notion precise later, see Sect. 5.3). The proof of this fact is an extension of ideas from [CPS09]. What we are able to show is that the integrand in (1.6) is an exact one form on our affine space of admissible perturbations regarded as a Banach manifold.

Theorem 1.3.3 *Let A_- be a self-adjoint operator and let B be a p-relative trace-class perturbation of A_-. We have that*

$$\lim_{t\to\infty} [-\left(\frac{t}{\pi}\right)^{1/2} \int_0^1 \operatorname{tr}\left(e^{-tA_s^2}(A_+ - A_-)\right)ds]$$

does not depend on a smooth path $\{A_s\}_{s\in[0,1]}$ in the affine space of p-relative trace-class perturbations of A_-, joining A_- and $A_- + B$.

The result above suggests that in the non-Fredholm case we can relate

$$\lim_{t\to\infty}[-\left(\frac{t}{\pi}\right)^{1/2}\int_0^1 \operatorname{tr}\left(e^{-tA_s^2}(A_+ - A_-)\right)ds]$$

to a generalisation of spectral flow along an admissible path joining $A_\pm$ when $A_\pm$ are unitarily equivalent. One consequence of our discussion in these notes is that when the Witten index exists so too does the limit on the right-hand side. Then by inserting kernel correction terms on the right-hand side we obtain a notion of a generalised spectral flow when the endpoints $A_\pm$ are not invertible.

Moreover under appropriate regularity assumptions for the respective spectral shift functions at zero we see that our generalised spectral flow can be expressed in terms of the spectral shift function just as in the Fredholm case. This fact may have applications in condensed matter theory where variations in parameters in the Hamiltonian can result in the closing of the spectral gap at zero so that one is forced to consider the non-Fredholm case.

1.4 Summary of the Exposition

In Chap. 2 we collect some preliminaries. A key technical tool that we employ, namely the theory of double operator integrals (DOI) is explained in this chapter. This is, loosely speaking, a differential calculus for functions of two non-commuting variables. In these notes the variables in question are operators on an infinite dimensional Hilbert space. In fact, in order that our text is more self contained, we give an exposition of the DOI method in Chap. 2. This technique has had many applications in non-commutative analysis in the last few years and we intend our exposition in these notes to be useful in other contexts.

Following this in Chap. 3 we give a detailed exposition of the setting for the recent results described here and explain the approximation scheme that we use in order to apply the earlier results of [Pus08, GLM+11] to prove them. The idea is to use a 'cut-off' that, in the case of differential operators, produces a sequence of pseudo-differential approximants. This is the technical heart of our method. The point is that for the approximants the principal trace formula is quickly established. Then in Chap. 6 we have to handle the convergence of both sides as we remove our cut-off thus proving the principal trace formula.

In Chap. 4 we first provide some expository discussion on the spectral shift function adapted to our setting. In particular we outline the perturbation determinant formulas for the spectral shift function. There is a subtle point that we need to be careful about in that for the pair $A_\pm$ the spectral shift function is a priori only defined up to a constant. Fixing this constant is essential but is not a trivial matter and we explain how to use the approximation scheme of the previous chapter to achieve that.

Next, in Chap. 5, comes our discussion of spectral flow from the analytic viewpoint. We do not start with the original approach of Atiyah-Patodi-Singer but instead explain the analytic approach introduced by John Phillips. We review analytic formulas for spectral flow prior to explaining how the spectral shift function may be used to compute spectral flow in some circumstances. The analytic formulas for spectral flow suggest a notion of a generalised spectral flow for paths of non-Fredholm operators and we explain some of its properties including a restricted homotopy invariance.

The preceding material is brought together in Chap. 6 by our exposition of the principal trace formula. The Chapter begins with a history of the principal trace formula followed immediately by its proof in both the heat kernel and resolvent formulations. Then we turn to some applications.

The first of these is a very general formula relating the spectral shift function for the pair $A_{\pm}$ to that for the pair $\boldsymbol{D}_A^* \boldsymbol{D}_A$, $\boldsymbol{D}_A \boldsymbol{D}_A^*$ that can be understood as having its roots in Pushnitski's work. This is followed by an exposition of the Witten index and a demonstration of its relation to the spectral shift function of the pair $\boldsymbol{D}_A^* \boldsymbol{D}_A$, $\boldsymbol{D}_A \boldsymbol{D}_A^*$. For completeness we explain how the homological index of [CK17] is connected to the Witten index. The introduction of the homological index enables us to discuss the anomaly (for which some background material is provided) and its relationship to the spectral shift function. Then our version of Pushnitski's formula provides an equality, via the principal trace formula, of the Witten index with a generalised spectral flow.

The final chapter is devoted to examples. The main results are about Dirac operators in arbitrary dimensions. We sketch some work on these operators that is contained in a companion monograph [CGL+22]. In particular we apply that to the regularity issue that arises for the spectral shift function at zero in connection with the existence of the Witten index. We also give a detailed one dimensional example of the general operator theoretic formalism developed earlier. In one dimension everything can be made explicit because the spectral shift function is comparatively easily computed. We see from this simple example that the anomaly, the spectral flow and the Witten index may be different (in contrast to the Fredholm case).

Chapter 2
Double Operator Integrals

In the present chapter we present the main technical tool of our approach: double operator integrals (DOIs). Double operator integrals are a tool for handling the differential calculus of multivariable functions of operators. This machinery is crucial in our approach and many aspects of the theory rely on this technique.

Double operator integrals first appeared in the work of Daletskii and Krein [DK56] who studied questions on the differentiability of operator-valued functions. The theory has been developed in the works of Birman and Solomyak in the series of papers [BS66, BS67] and [BS73]. In this chapter we present a short overview of the theory of double operator integrals and their properties. Note that the material presented here does not represent a full review of the topic of double operator integrals as we concentrate on the specifics of the theory relevant to the main objectives of the present notes. We refer the reader to the surveys [BS03, Pel16, ST19] for more detailed expositions.

For the convenience of the reader we explain the idea behind double operator integrals in the special case when the double operator integral is generated by two self-adjoint operators with purely discrete spectra. In this case, the construction of double operator integrals is significantly simpler.

The crucial results of this chapter are in Sect. 2.4. One of these results (see Proposition 2.4.6) may be viewed as a non-commutative version of the Fundamental Theorem of Calculus. Another result, is the approximation result, proved in [CGL+16a], for operator valued functions in the trace-class ideal suitable for Dirac operators. The latter result allows us to implement the approximation technique introduced later in Chap. 3.

A. Carey, G. Levitina, *Index Theory Beyond the Fredholm Case*, Lecture Notes in Mathematics 2323, https://doi.org/10.1007/978-3-031-19436-8_2

2.1 Double Operator Integrals in the Discrete Setting

In this section we present the construction and properties of double operator integrals in the discrete case. That is, we assume that the operators generating a double operator integral have purely discrete spectrum. In this case, the definition of double operator integrals is simple and the main properties can be explained without going into technicalities, which are unavoidable for operators with continuous spectrum. In this section we discuss double operator integrals defined on the Hilbert-Schmidt class $\mathcal{L}_2(\mathcal{H})$ only. As shown in the next section, double operator integrals can be defined as mappings on $\mathcal{L}(\mathcal{H})$. This, however, requires additional technicalities, which we wish to avoid in this introductory discussion.

Throughout this section we assume that A, B are self-adjoint operators with purely discrete spectrum, that is spectra of A and B consist of isolated eigenvalues of finite multiplicities. Denote by $\{\xi_k\}_{k=1}^\infty$ and $\{\eta_k\}_{k=1}^\infty$ the orthonormal bases of eigenvectors and by $\{\lambda_k\}_{k=1}^\infty$ and $\{\mu_k\}_{k=1}^\infty$ the eigenvalues of A and B, respectively, counting multiplicities.

Assume that φ is a function on $\mathbb{R}^2$, such that $\alpha = \sup_{j,k=1,\dots,\infty} |\varphi(\lambda_j, \mu_k)| < \infty$ and let $X \in \mathcal{L}_2(\mathcal{H})$. Denote by P_e the orthonormal projection on the subspace generated by $e \in \mathcal{H}$. Since $\{\xi_k\}_{k=1}^\infty$ and $\{\eta_k\}_{k=1}^\infty$ are orthonormal bases in $\mathcal{H}$, for the operator $\sum_{j=1}^\infty \sum_{k=1}^\infty \varphi(\lambda_j, \mu_k) P_{\xi_j} X P_{\eta_k}$ (where the series converges in the strong operator topology), we have that

$$\begin{aligned}
\Big\| \sum_{j=1}^\infty \sum_{k=1}^\infty \varphi(\lambda_j, \mu_k) P_{\xi_j} X P_{\eta_k} \Big\|_2^2 &= \sum_{i=1}^\infty \Big\| \sum_{j=1}^\infty \sum_{k=1}^\infty \varphi(\lambda_j, \mu_k) P_{\xi_j} X P_{\eta_k} \eta_i \Big\|^2 \\
&= \sum_{i=1}^\infty \Big\| \sum_{j=1}^\infty \varphi(\lambda_j, \mu_i) P_{\xi_j} X \eta_i \Big\|^2 = \sum_{i=1}^\infty \sum_{\ell=1}^\infty \Big| \langle \sum_{j=1}^\infty \varphi(\lambda_j, \mu_i) P_{\xi_j} X \eta_i, \xi_\ell \rangle \Big|^2 \\
&= \sum_{i,\ell=1}^\infty |\varphi(\lambda_\ell, \mu_i) \langle X\eta_i, \xi_\ell \rangle|^2 \\
&\le \alpha^2 \sum_{i,\ell=1}^\infty |\langle X\eta_i, \xi_\ell \rangle|^2 = \alpha^2 \sum_{i=1}^\infty \|X\eta_i\|^2 = \alpha^2 \|X\|_2^2.
\end{aligned}$$

Thus, for every $X \in \mathcal{L}_2(\mathcal{H})$ the operator $\sum_{j=1}^\infty \sum_{k=1}^\infty \varphi(\lambda_j, \mu_k) P_{\xi_j} X P_{\eta_k}$ is in the Hilbert-Schmidt class $\mathcal{L}_2(\mathcal{H})$ too.

The above discussion allows us to define the discrete *double operator integral* $T_\varphi^{A,B}$ as a bounded mapping $T_\varphi^{A,B} : \mathcal{L}_2(\mathcal{H}) \to \mathcal{L}_2(\mathcal{H})$, defined via

$$T_\varphi^{A,B}(X) = \sum_{j=1}^\infty \sum_{k=1}^\infty \varphi(\lambda_j, \mu_k) P_{\xi_j} X P_{\eta_k}, \quad X \in \mathcal{L}_2(\mathcal{H}). \tag{2.1}$$

The function φ is called the *symbol* of the double operator integral $T_\varphi^{A,B}$ and as shown above

$$\|T_\varphi^{A,B}\|_{\mathcal{L}(\mathcal{L}_2(\mathcal{H}))} \le \sup\{|\varphi(\lambda_j, \mu_k)|, \lambda_j \in \sigma(A), \mu_k \in \sigma(B)\}. \tag{2.2}$$

The double operator integral can be equivalently rewritten in terms of spectral measures of A and B. Denote by $\{\lambda_k\}_{k=1}^\infty$ and $\{\mu_k\}_{k=1}^\infty$ the sets of distinct eigenvalues of A and B, respectively. Then, by spectral theory we have that

$$E_A(\{\lambda_k\}) = \sum_{i:\lambda_i=\lambda_k} P_{\xi_i}, \quad k = 1, \ldots \infty,$$

and

$$E_B(\{\mu_k\}) = \sum_{i:\mu_i=\mu_k} P_{\eta_i}, \quad k = 1, \ldots \infty.$$

Therefore, the definition (2.1) can be rewritten as

$$T_\varphi^{A,B}(X) = \sum_{j=1}^\infty \sum_{k=1}^\infty \varphi(\lambda_j, \mu_k) E_A(\{\lambda_j\}) X E_B(\{\mu_k\}).$$

In the case when A, B are unitary operators the double operator integral $T_\varphi^{A,B}$ can be defined similarly, with the only difference that the function φ is defined on the torus $\mathbb{T}$. Further in this section we distinguish the self-adjoint and unitary cases only if it is stated explicitly.

When it is clear from the context which operators are being used to define a double operator integral, the notation T_φ is used instead of $T_\varphi^{A,B}$.

Note that the double operator integral $T_\varphi^{A,B}(X)$ is simply the Schur product $\{\varphi(\lambda_i, \mu_j)\}_{i,j} * X$ of the matrices $\{\varphi(\lambda_i, \mu_j)\}_{i,j}$ and $X = \{x_{ij}\}_{i,j}$ (see e.g. [Bha96, HJ12]). Indeed, for any $1 \le j_0, k_0 < \infty$ we have that

$$\begin{aligned}\langle T_\varphi^{A,B}(X)\eta_{k_0}, \xi_{j_0}\rangle &= \sum_{j=1}^\infty \varphi(\lambda_j, \mu_{k_0})\langle P_{\xi_j} X \eta_{k_0}, \xi_{j_0}\rangle \\ &= \varphi(\lambda_{j_0}, \mu_{k_0})\langle X\eta_{k_0}, \xi_{j_0}\rangle = \varphi(\lambda_{j_0}, \mu_{k_0}) x_{j_0 k_0},\end{aligned}$$

where

$$x_{ij} = \langle X(\eta_j), \xi_i\rangle, \quad 1 \le i, j < \infty$$

is the representation of an operator $X \in \mathcal{L}_2(\mathcal{H})$ as a matrix with respect to $\{\xi_i\}$ and $\{\eta_j\}$.

In particular, double operator integrals are considered as continuous versions of Schur multipliers. In the general case, when the operators A and B are arbitrary self-adjoint operators on $\mathcal{H}$, the double operator integral $T_\varphi^{A,B}$ can be also understood as a version of a Schur multiplier by considering the decomposition of the Hilbert space $\mathcal{H}$ as a direct integral with respect to the spectral measures of A and B (see e.g. [BS03, Section 3.2]).

It is clear from the definition that double operator integrals are linear with respect to the symbol, that is

$$T_{\alpha\varphi+\beta\psi} = \alpha T_\varphi + \beta T_\psi, \alpha, \beta \in \mathbb{C}.$$

Furthermore, it is easy to see that $T_{\psi\varphi} = T_\varphi T_\psi$ and $T_\varphi = 1$ provided that $\varphi = 1$.

The importance of double operator integrals in perturbation theory in general and in index theory in particular stems from the so-called perturbation formula, which was first discovered by Löwner [L3̈4]. Let, as before, A, B be self-adjoint operators with purely discrete spectra and assume, in addition, that $A - B \in \mathcal{L}_2(\mathcal{H})$. For a differentiable function f we define the divided difference $f^{[1]}$ by

$$f^{[1]}(\lambda, \mu) = \begin{cases} \frac{f(\lambda)-f(\mu)}{\lambda-\mu}, \lambda \neq \mu \\ f'(\lambda), \lambda = \mu. \end{cases}$$

Then

$$f(A) - f(B) = T_{f^{[1]}}^{A,B}(A - B). \tag{2.3}$$

Indeed, by the spectral theorem we have that $A = \sum_{j=1}^\infty \lambda_j P_{\xi_j}$ and $B = \sum_{k=1}^\infty \mu_k P_{\eta_k}$. Therefore, $f(A)\xi_j = f(\lambda_j)P_{\xi_j}$ and $f(B)\eta_k = f(\mu_k)P_{\eta_k}$ for any $j, k = 1, \dots, \infty$ and so for $j, k = 1, \dots, \infty$ such that $\lambda_j \neq \mu_k$ we have

$$\begin{aligned} \langle (f(A) - f(B))\eta_k, \xi_j\rangle &= (f(\lambda_i) - f(\mu_k))\langle \eta_k, \xi_j\rangle \\ &= \frac{f(\lambda_i) - f(\mu_k)}{\lambda_j - \mu_k}\langle (A - B)\eta_k, \xi_j\rangle = \langle T_{f^{[1]}}^{A,B}(A - B)\eta_k, \xi_j\rangle. \end{aligned}$$

Clearly, if $\lambda_j = \mu_k$, then $\langle (f(A) - f(B))\eta_k, \xi_j\rangle = 0 = \langle T_{f^{[1]}}^{A,B}(A - B)\eta_k, \xi_j\rangle$. In particular, combining (2.2) and (2.3) we have that

$$\|f(A) - f(B)\|_2 \leq \|T_{f^{[1]}}\|_{\mathcal{L}(\mathcal{L}_2(\mathcal{H}))}\|A - B\|_2 = \|f\|_{Lip}\|A - B\|_2, \tag{2.4}$$

where $\|f\|_{Lip}$ is the Lipschitz norm of the function f.

In applications, relevant to the present lecture notes, the operator A is typically a differential operator and B is its perturbation by a potential function. In particular, the difference $A - B$ is never a compact operator and so estimates in operator ideals similar to (2.4) do not hold. Nevertheless for specific choice of functions f the difference $f(A) - f(B)$ can be trace-class. In this case, the Eq. (2.3) is modified to express the difference $f(A) - f(B)$ as a double operator applied to the difference of resolvents $(A-i)^{-1} - (B-i)^{-1}$ (or higher powers of resolvents) which may be a trace-class operator.

To demonstrate the idea, let, as before, A and B be self-adjoint operators with purely discrete spectra and assume that $(A-i)^{-1} - (B-i)^{-1}$ is a Hilbert-Schmidt class operator. Let $\gamma(\lambda) = \frac{\lambda+i}{\lambda-i}$, $\lambda \in \mathbb{R}$, denote the Cayley transform and let f be a differentiable function on $\mathbb{R}$. Consider the double operator integral $T_\psi^{\gamma(A),\gamma(B)}$, where $\psi = (f \circ \gamma^{-1})^{[1]}$. Then, by (2.3) we have

$$T_\psi^{\gamma(A),\gamma(B)}\Big((A-i)^{-1} - (B-i)^{-1}\Big) = \frac{1}{2i} T_\psi^{\gamma(A),\gamma(B)}\Big(\gamma(A) - \gamma(B)\Big) = \frac{1}{2i}(f(A) - f(B)).$$

In particular, if f is a sufficiently nice function to generate a bounded (on $\mathcal{L}_2(\mathcal{H})$) double operator integral $T_\psi^{\gamma(A),\gamma(B)}$, then we have the estimate

$$\|f(A) - f(B)\|_2 \leq 2\|T_\psi^{\gamma(A),\gamma(B)}\|_{\mathcal{L}(\mathcal{L}_2(\mathcal{H}))}\|(A-i)^{-1} - (B-i)^{-1}\|_2. \tag{2.5}$$

Estimates analogous to (2.5) in the present lecture notes are important in the case of the trace-class norm. These allow us to establish that relevant operators in the principal trace formula (1.2) are trace-class. In the Sect. 2.3 we will present the details of a construction due to D. Yafaev which allows to write $f(A) - f(B)$ as a double operator integral applied to the difference of higher powers of resolvents $(A-ai)^{-m} - (B-ai)^{-m}$, $m \in \mathbb{N}$, $a \in \mathbb{R}$.

2.2 Double Operator Integrals in the General Setting

In this section we discuss the construction of double operator integrals in the general setting. We refer the reader to [ST19] and [Pel16] to a full exposition and the relevant proofs.

Double operator integrals appeared first in the paper [DK56] for a restrictive class of functions. The general theory of double operator integrals was developed in [BS66, BS67, BS73] and there are several available constructions (see e.g. [dPWS02, Pel85, PSS13, ACDS09]). In these lecture notes we use double operator integrals as operators on $\mathcal{L}_1(\mathcal{H})$ only. However, to describe the trace-class setting we need to first handle double operator integrals on the Hilbert-Schmidt ideal.

To begin we recall here the classical definition due to Birman and Solomyak [BS66]. Suppose that A, B are self-adjoint operators on a Hilbert space $\mathcal{H}$. Denote by E_A and E_B the ($\mathcal{L}(\mathcal{H})$-valued) spectral measures on $\mathbb{R}$ of A and B, respectively. Consider the $\mathcal{L}(\mathcal{L}_2(\mathcal{H}))$-valued measures on $\mathbb{R}$ defined by

$$\mathcal{E}(\sigma_1) : X \to E_A(\sigma_1)X,$$
$$\mathcal{F}(\sigma_2) : X \to X E_B(\sigma_2), \quad X \in \mathcal{L}_2(\mathcal{H}),$$

where σ_1, σ_2 are Borel sets in $\mathbb{R}$. It is clear that the $\mathcal{E}$ and $\mathcal{F}$ are commuting spectral measures on $\mathbb{R}$ in the sense that $\mathcal{E}(\sigma_1)\mathcal{F}(\sigma_2) = \mathcal{F}(\sigma_2)\mathcal{E}(\sigma_1)$ for all Borel sets $\sigma_1\sigma_2$ in $\mathbb{R}$. Define the product of two measures $\mathcal{E}$ and $\mathcal{F}$

$$\nu(\sigma_1 \times \sigma_2) = \mathcal{E}(\sigma_1)\mathcal{F}(\sigma_2),$$

that is

$$\nu(\sigma_1 \times \sigma_2)(X) = E_A(\sigma_1)XE_B(\sigma_2), \quad X \in \mathcal{L}_2(\mathcal{H}).$$

It is proved by Birman and Solomyak [BS96, Theorem 2] (see also [ST19, Proposition 2.5.1]) that this is a countably additive (in the strong operator topology) projection-valued measure on $\mathbb{R}^2$. Denote by $L_\infty(\mathbb{R}^2; \nu)$ the space all ν-measurable functions ψ on $\mathbb{R}^2$ such that

$$\|\psi\|_\infty = \nu\text{-sup}\,\psi = \inf\{\alpha \geq 0 : |\psi| \leq \alpha \ \nu - \text{a.e.}\} < \infty$$

(see [BS87, Section 5.1] for precise definitions).

Definition 2.2.1 For a function $\psi \in L_\infty(\mathbb{R}^2; \nu)$ the (Birman-Solomyak) double operator integral $T_\psi^{A,B} : \mathcal{L}_2(\mathcal{H}) \to \mathcal{L}_2(\mathcal{H})$ (with the symbol ψ built over spectral measures of A and B) is defined as the integral of the function ψ with respect to the spectral measure ν, that is

$$T_\psi^{A,B}(X) := \int_{\mathbb{R}^2} \psi(\omega)d\nu(\omega)(X), \quad X \in \mathcal{L}_2(\mathcal{H}).$$

The other frequently used notation is

$$T_\psi^{A,B}(X) = \int_{\mathbb{R}^2} \psi(\lambda, \mu)dE_A(\lambda)XdE_B(\mu), \quad X \in \mathcal{L}_2(\mathcal{H}).$$

By spectral theory, the condition $\psi \in L_\infty(\mathbb{R}^2; \nu)$ is necessary and sufficient for boundedness of the operator $T_\psi^{A,B} : \mathcal{L}_2(\mathcal{H}) \to \mathcal{L}_2(\mathcal{H})$ [BS66], [ST19, Proposition 3.2.2].

Remark 2.2.2

(i) For a DOI $T_\psi^{A,B}$ with symbol ψ the values of ψ outside some Borel subset $\Omega \subset \mathbb{R}^2$ containing $\sigma(A) \times \sigma(B)$ are inessential. Namely, if $\sigma(A) \cup \sigma(B) \subset \Omega$ and $\psi|_\Omega$ denotes the restriction of ψ onto Ω, then

$$T_\psi^{A,B} = T_{\psi|_\Omega}^{A,B}.$$

(ii) One can also introduce the double operator integral $T_\psi^{U,V}$ for a pair of normal operators U, V.

◇

Next, we discuss double operator integrals on $\mathcal{L}(\mathcal{H})$. Recall that $\mathcal{L}(\mathcal{H})$ is dual to $\mathcal{L}_1(\mathcal{H})$ via trace duality given by

$$\langle T, S\rangle = \operatorname{tr}(TS^*), \quad T \in \mathcal{L}_1(\mathcal{H}), S \in \mathcal{L}(\mathcal{H}).$$

Since $\mathcal{L}_1(\mathcal{H}) \subset \mathcal{L}_2(\mathcal{H})$, one can easily conclude that $T_\psi^{A,B}(X) \in \mathcal{L}_2(\mathcal{H})$ for any $X \in \mathcal{L}_1(\mathcal{H})$ and $\psi \in L_\infty(\mathbb{R}^2; \nu)$. Assume that ψ is such that the double operator integral $T_\psi^{B,A}$ is bounded on $\mathcal{L}_1(\mathcal{H})$, then $T_{\bar{\psi}}^{B,A}$ is a bounded operator on $\mathcal{L}(\mathcal{H})$. Therefore, one can define the double operator integral $T_\psi^{A,B}$ on $\mathcal{L}(\mathcal{H})$ by duality

$$T_\psi^{A,B}(T) = (T_{\bar{\psi}}^{B,A})^*(T), \quad T \in \mathcal{L}(\mathcal{H}). \tag{2.6}$$

We introduce

$$\mathfrak{M}_1 := \{\psi \in L_\infty(\mathbb{R}^2; \nu) \mid T_\psi^{A,B} \in \mathcal{L}(\mathcal{L}_1(\mathcal{H}))\},$$
$$\mathfrak{M}_\infty := \{\psi \in L_\infty(\mathbb{R}^2; \nu) \mid T_\psi^{A,B} \in \mathcal{L}(\mathcal{L}(\mathcal{H}))\}.$$

In addition, we set

$$\|\psi\|_{\mathfrak{M}_1} := \|T_\psi^{A,B}\|_{\mathcal{L}(\mathcal{L}_1(\mathcal{H}))}, \quad \|\psi\|_{\mathfrak{M}_\infty} := \|T_\psi^{A,B}\|_{\mathcal{L}(\mathcal{L}(\mathcal{H}))}.$$

It follows from the definition that

$$\mathfrak{M} := \mathfrak{M}_1 = \mathfrak{M}_\infty,$$

and

$$\|\psi\|_{\mathfrak{M}} := \|\psi\|_{\mathfrak{M}_1} = \|\psi\|_{\mathfrak{M}_\infty}, \quad \psi \in \mathfrak{M}.$$

Thus, the definition of the double operator integral $T_\psi^{A,B}$ on $\mathcal{L}(\mathcal{H})$ heavily relies on the fact that $T_{\bar{\psi}}^{A,B}$ is a bounded operator on $\mathcal{L}_1(\mathcal{H})$. However, in contrast to the double operator integrals on $\mathcal{L}_2(\mathcal{H})$, the condition that $\psi \in L_\infty(\mathbb{R}^2; \nu)$ does not guarantee that $T_\psi^{A,B} \in \mathcal{L}(\mathcal{L}_1(\mathcal{H}))$.

Nevertheless, the class of symbols ψ generating bounded (on $\mathcal{L}(\mathcal{H})$) double operator integral admits a useful characterisation based on separation of variables. This approach was introduced by M. Birman and M. Solomyak in [BS66] and developed further in [BS67, BS73] and [Pel85]. We will next briefly discuss this approach.

We firstly mention some of the basic properties of double operator integrals.

Proposition 2.2.3 *Let A, B be arbitrary self-adjoint operators on $\mathcal{H}$ with common dense domain. Suppose that $\psi_1, \psi_2, \psi \in \mathfrak{M}$. We have*

(i) $T^{A,B}_{\psi_1+\psi_2} = T^{A,B}_{\psi_1} + T^{A,B}_{\psi_2}$;
(ii) $T^{A,B}_{\psi_1\psi_2} = T^{A,B}_{\psi_1} \circ T^{A,B}_{\psi_2}$;
(iii) if $\psi(t_1, t_2) = h_1(t_1)h_2(t_2)$, *for some bounded functions h_1 and h_2 on $\mathbb{R}$, then* $T^{A,B}_{\psi}(X) = h_1(A)Xh_2(B)$;

Proof Parts (i) and (ii) follow directly from spectral theory.
(iii) Note that

$$T^{A,B}_{h_1}(X) = \int_{\mathbb{R}^2} h_1(\omega_1)d\nu(\omega)(X) = \int_{\mathbb{R}} h_1(\omega_1)dE_A(\omega_1) \cdot X = h_1(A)X$$

and, similarly,

$$T^{A,B}_{h_2}(X) = Xh_2(B).$$

Referring to part (ii) we infer that

$$T^{A,B}_{\psi}(X) = (T^{A,B}_{h_1} \circ T^{A,B}_{h_2})(X) = T^{A,B}_{h_1}(Xh_2(B)) = h_1(A)Xh_2(B).$$

□

Proposition 2.2.3 in particular implies that if a function ψ admits a representation

$$\psi(\lambda, \mu) = \sum_{n\in\mathbb{N}} \alpha_n(\lambda)\beta_n(\mu),$$

with $\alpha_n \in L_\infty(\mathbb{R}, dE_A)$, $\beta_n \in L_\infty(\mathbb{R}, dE_B)$ and

$$\sum_{n\in\mathbb{N}} \|\alpha_n\|_\infty \|\beta_n\|_\infty < \infty,$$

(that is, ψ belongs to the projective tensor product of $L_\infty(\mathbb{R}, dE_A)$ and $L_\infty(\mathbb{R}, dE_B)$), then

$$T^{A,B}_{\psi}(X) = \sum_{n\in\mathbb{N}} \Big(\int_{\mathbb{R}} \alpha_n(\lambda)dE_A(\lambda)\Big) \cdot X \cdot \Big(\int_{\mathbb{R}} \beta_n(\mu)dE_B(\mu)\Big).$$

In [Pel85] V. Peller has extended this idea to the integral tensor product $L_\infty(\mathbb{R}, dE_A)\hat{\otimes}_i L_\infty(\mathbb{R}, dE_B)$ of bounded functions. The class $L_\infty(\mathbb{R}, dE_A)\hat{\otimes}_i L_\infty(\mathbb{R}, dE_B)$ consists of all functions $\psi : \mathbb{R}^2 \to \mathbb{C}$ admitting a representation

$$\psi(\lambda, \mu) = \int_{\Omega} \alpha(\lambda, t)\beta(\mu, t)\, d\eta(t), \quad (\lambda, \mu) \in \mathbb{R}^2, \tag{2.7}$$

where $(\Omega, d\eta(t))$ is an auxiliary σ-finite measure space and

$$\int_\Omega \|\alpha(\cdot,t)\|_\infty \|\beta(\cdot,t)\|_\infty d\eta(t) < \infty.$$

The class $L_\infty(\mathbb{R}, dE_A)\hat{\otimes}_i L_\infty(\mathbb{R}, dE_B)$ is an algebra with respect to pointwise addition and multiplication [ACDS09, Proposition 4.10] and the formula

$$\|\psi\|_{L_\infty(\mathbb{R},dE_A)\hat{\otimes}_i L_\infty(\mathbb{R},dE_B)} = \inf \int_\Omega \|\alpha(\cdot,t)\|_\infty \|\beta(\cdot,t)\|_\infty d\eta(t) < \infty,$$

where the infimum is taken over all possible representations of ψ of the form (2.7), defines a norm on the space $L_\infty(\mathbb{R}, dE_A)\hat{\otimes}_i L_\infty(\mathbb{R}, dE_B)$.

Assume that $\psi \in L_\infty(\mathbb{R}, dE_A)\hat{\otimes}_i L_\infty(\mathbb{R}, dE_B)$ with a representation of the form (2.7). It is clear that the function

$$t \mapsto \Big(\int_{\mathbb{R}} \alpha(\lambda,t)dE_A(\lambda)\Big)X\Big(\int_{\mathbb{R}} \beta(\mu,t)dE_B(\mu)\Big)$$

is weakly measurable and

$$\int_\Omega \Big\|\Big(\int_{\mathbb{R}} \alpha(\lambda,t)dE_A(\lambda)\Big)X\Big(\int_{\mathbb{R}} \beta(\mu,t)dE_B(\mu)\Big)\Big\| d\eta(t) < \infty.$$

In particular, for every $y \in \mathcal{H}$ the integral

$$\int_\Omega \big(\alpha(A,t)X\beta(B,t)\big)(y)d\eta(t)$$

is a well-defined Bochner integral. This allows us to define the integral $\mathcal{T}(X) = \int_\Omega \alpha(A,t)X\beta(B,t)d\eta(t)$ as follows

$$\Big(\int_\Omega \alpha(A,t)X\beta(B,t)d\eta(t)\Big)(y) = \int_\Omega \big(\alpha(A,t)X\beta(B,t)\big)(y)d\eta(t),$$

with

$$\|\mathcal{T}(X)\| \le \|\psi\|_{L_\infty(\mathbb{R},dE_A)\hat{\otimes}_i L_\infty(\mathbb{R},dE_B)} \|X\|. \tag{2.8}$$

As shown in [ACDS09, Lemma 4.3], the bounded operator $\mathcal{T} : \mathcal{L}(\mathcal{H}) \to \mathcal{L}(\mathcal{H})$ is well-defined in the sense that it does not depend on a particular representation of the function ψ of the form (2.7). Furthermore, similarly to the proof of Proposition 2.2.3 one can show that

$$T_\psi^{A,B}(X) = \int_\Omega \alpha(A,t)X\beta(B,t)d\eta(t). \tag{2.9}$$

Thus, if $\psi \in L_\infty(\mathbb{R}, dE_A)\hat{\otimes}_i L_\infty(\mathbb{R}, dE_B)$, then the double operator integral $T_\psi^{A,B}$ can be defined via (2.9) and, furthermore, by (2.8) is bounded on $\mathcal{L}(\mathcal{H})$, that is $\psi \in \mathfrak{M}$. V. Peller showed in [Pel85] that the condition $\psi \in L_\infty(\mathbb{R}, dE_A)\hat{\otimes}_i L_\infty(\mathbb{R}, dE_B)$ is also necessary for the inclusion $\psi \in \mathfrak{M}$.

Namely, we have the following result.

Theorem 2.2.4 ([Pel85] (see also [BS03, Theorem 4.1], [Pel16, Theorem 1.1.1.])) *The following conditions are equivalent:*

(i) $\psi \in \mathfrak{M}$.
(ii) $\psi \in L_\infty(\mathbb{R}, dE_A)\hat{\otimes}_i L_\infty(\mathbb{R}, dE_B)$*, in which case* $\|\psi\|_{\mathfrak{M}} \le \|\psi\|_{L_\infty(\mathbb{R},dE_A)\hat{\otimes}_i L_\infty(\mathbb{R},dE_B)}$.
(iii) The function $\psi(\cdot, \cdot)$ *admits a representation of the form*

$$\psi(\lambda, \mu) = \int_\Omega \alpha(\lambda, t)\beta(\mu, t)\, d\eta(t), \quad (\lambda, \mu) \in \mathbb{R}^2,$$

where $(\Omega, d\eta(t))$ *is an auxiliary measure space and*

$$C_\alpha^2 := \sup_{\lambda\in\mathbb{R}} \int_\Omega |\alpha(\lambda, t)|^2\, d\eta(t) < \infty, \quad C_\beta^2 := \sup_{\mu\in\mathbb{R}} \int_\Omega |\beta(\mu, t)|^2\, d\eta(t) < \infty.$$

$$\|\psi\|_{\mathfrak{M}} \le C_\alpha C_\beta.$$

The integral representation (2.9) via separation of variables implies the following simple property of double operator integrals.

Proposition 2.2.5 *Let* A, B *be arbitrary self-adjoint operators on* $\mathcal{H}$ *with common dense domain and let* $\psi \in \mathfrak{M}$. *If* $\psi(t_1, t_2) = \psi(t_2, t_1)$ *for all* $t_1, t_2 \in \mathbb{R}$*, then* $T_\psi^{A,A}(1) = f(A)$*, where* $f(t) = \psi(t, t)$*;*

Proof By Theorem 2.2.4 we can find a measure space $(\Omega, d\eta(t))$ such that

$$\psi(\lambda, \mu) = \int_\Omega \alpha(\lambda, t)\beta(\mu, t)d\eta(t), \quad (\lambda, \mu) \in \mathbb{R}^2,$$

with $\int_\Omega \|\alpha(\cdot, t)\|_\infty \|\beta(\cdot, t)\|_\infty d\eta(t) < \infty$. For any $X \in \mathcal{L}(\mathcal{H})$ we have

$$\begin{aligned} T_\psi^{A,A}(X) &= \int_{\mathbb{R}^2} \psi(\lambda, \mu) dE_A(\lambda) X dE_A(\mu) \\ &= \int_\Omega \Big(\int_\mathbb{R} \alpha(\lambda, t) dE_A(\lambda)\Big) X \Big(\int_\mathbb{R} \beta(\mu, t) dE_A(\mu)\Big) d\eta(t). \end{aligned}$$

Therefore,

$$
\begin{aligned}
T_\psi^{A,A}(1) &= \int_\Omega \Big(\int_{\mathbb{R}} \alpha(\lambda,t)dE_A(\lambda)\Big)\cdot\Big(\int_{\mathbb{R}} \beta(\mu,t)dE_A(\mu)\Big)d\eta(t)\\
&= \int_\Omega \Big(\int_{\mathbb{R}} \alpha(\lambda,t)\beta(\lambda,t)dE_A(\lambda)\Big)d\eta(t)\\
&= \int_{\mathbb{R}} \Big(\int_\Omega \alpha(\lambda,t)\beta(\lambda,t)d\eta(t)\Big)dE_A(\lambda) = \int_{\mathbb{R}} \psi(\lambda,\lambda)dE_A(\lambda) = f(A).
\end{aligned}
$$

□

The representation of double operator integrals via separation of variables easily implies the following change of variables formula for double operator integrals.

Proposition 2.2.6 *Let A, B be arbitrary self-adjoint operators on $\mathcal{H}$ with common dense domain and let $\psi \in \mathfrak{M}$. If $f, g : \mathbb{R} \to \mathbb{R}$ are Borel functions, then $T_\psi^{f(A),g(B)} = T_{\tilde{\psi}}^{A,B}$, where $\tilde{\psi}(\lambda,\mu) = \psi(f(\lambda), g(\mu))$, $\lambda, \mu \in \mathbb{R}$.*

Proof By spectral theory, the spectral measure $E_{f(A)}$ of the operator $f(A)$ is given by $E_{f(A)}(M) = E_A(f^{-1}(M))$ for any Borel set $M \subset \mathbb{R}$. In particular, for any Borel function $h : \mathbb{R} \to \mathbb{R}$ we have that $\int_{\mathbb{R}} h(\lambda)dE_{f(A)}(\lambda) = \int_{\mathbb{R}}(h \circ f)(\lambda)dE_A(\lambda)$. Similarly, $\int_{\mathbb{R}} h(\lambda)dE_{g(B)}(\lambda) = \int_{\mathbb{R}}(h \circ g)(\lambda)dE_B(\lambda)$.

As before, we can find a measure space $(\Omega, d\eta(t))$ such that

$$
\psi(\lambda,\mu) = \int_\Omega \alpha(\lambda,t)\beta(\mu,t)d\eta(t), \quad (\lambda,\mu) \in \mathbb{R}^2,
$$

with $\int_\Omega \|\alpha(\cdot,t)\|_\infty \|\beta(\cdot,t)\|_\infty d\eta(t) < \infty$. Then $\tilde{\psi}$ has representation

$$
\tilde{\psi}(\lambda,\mu) = \int_\Omega \alpha(f(\lambda),t)\beta(g(\mu),t)d\eta(t), \quad (\lambda,\mu) \in \mathbb{R}^2.
$$

Hence, for any $X \in \mathcal{L}(\mathcal{H})$ we have

$$
\begin{aligned}
T_\psi^{f(A),g(B)}(X) &= \int_\Omega \Big(\int_{\mathbb{R}} \alpha(\lambda,t)dE_{f(A)}(\lambda)\Big)X\Big(\int_{\mathbb{R}} \beta(\mu,t)dE_{g(B)}(\lambda)\Big)d\eta(t)\\
&= \int_\Omega \Big(\int_{\mathbb{R}} \alpha(f(\lambda),t)dE_A(\lambda)\Big)X\Big(\int_{\mathbb{R}} \beta(g(\mu),t)dE_B(\mu)\Big)d\eta(t)\\
&= T_{\tilde{\psi}}^{A,B}(X).
\end{aligned}
$$

□

Remark 2.2.7 The change of variables formula of Proposition 2.2.6 also holds in the case, when $f(A), g(B)$ are unitaries. ◇

Next we discuss one particular class of symbols of double operator integrals, which will be used further in the rest of this chapter.

Theorem 2.2.8 ([BS03, Theorem 5.2]) *Suppose that there exist* $0 \le m_1 < 1$ *and* $1 < m_2$ *such that*

$$\sup_{\mu\in\mathbb{R}} \int_{\mathbb{R}} \left(|\xi|^{m_1} + |\xi|^{m_2}\right) \left|\widehat{\psi}(\xi,\mu)\right|^2 d\xi = C_0^2 < \infty,$$

where $\widehat{\psi}(\xi,\mu)$ *stands for the partial Fourier transform of* ψ *with respect to the first variable,*

$$\widehat{\psi}(\xi,\mu) = (2\pi)^{-1} \int_{\mathbb{R}} \psi(\lambda,\mu) e^{-i\xi\lambda}\, d\lambda, \quad (\xi,\mu) \in \mathbb{R}^2.$$

Then $\psi \in \mathfrak{M}$ *and*

$$\|\psi\|_{\mathfrak{M}} \le \text{const}\, C_0,$$

where the constant depends on m_1 *and* m_2 *only.*

Proof Since $m_1 < 1 < m_2$, we have that

$$\int_{\mathbb{R}} \left(|\xi|^{m_1} + |\xi|^{m_2}\right)^{-1} d\xi = 2 \int_0^{+\infty} \frac{dr}{|r|^{m_1} + |r|^{m_2}} =: C^2 \in (0,\infty). \tag{2.10}$$

That is, $f_{m_1,m_2}(\xi) = (|\xi|^{m_1} + |\xi|^{m_2})^{-1/2}$, $m_1 < 1 < m_2$, is square-integrable (with respect to Lebesgue measure on $\mathbb{R}$). Therefore, by the assumption and Hölder's inequality, we infer that

$$\begin{aligned}
\int_{\mathbb{R}} \left|\widehat{\psi}(\xi,\mu)\right| d\xi &= \int_{\mathbb{R}} \left[\left(|\xi|^{m_1} + |\xi|^{m_2}\right)^{\frac{1}{2}} \left|\widehat{\psi}(\xi,\mu)\right|\right] \left(|\xi|^{m_1} + |\xi|^{m_2}\right)^{-\frac{1}{2}} d\xi \\
&\le \left(\int_{\mathbb{R}} \left[\left(|\xi|^{m_1} + |\xi|^{m_2}\right)^{\frac{1}{2}} \left|\widehat{\psi}(\xi,\mu)\right|\right]^2 d\xi\right)^{1/2} \left(\int_{\mathbb{R}} \left(|\xi|^{m_1} + |\xi|^{m_2}\right)^{-1} d\xi\right)^{1/2} \\
&\le C_0 \left(\int_{\mathbb{R}} \left(|\xi|^{m_1} + |\xi|^{m_2}\right)^{-1} d\xi\right)^{1/2} \overset{(2.10)}{=} C_0\, C
\end{aligned}$$

uniformly for $\mu \in \mathbb{R}$. Hence, $\widehat{\psi}(\cdot,\mu) \in L_1(\mathbb{R})$ and $\sup_{\mu\in\mathbb{R}} \left\|\widehat{\psi}(\cdot,\mu)\right\|_{L_1(\mathbb{R})} < \infty$. By the inverse Fourier transform theorem

$$\begin{aligned}
\psi(\lambda,\mu) &= \int_{\mathbb{R}} \widehat{\psi}(\xi,\mu)\, e^{i\xi\lambda}\, d\xi \\
&= \int_{\mathbb{R}} e^{i\xi\lambda} \left(|\xi|^{m_1} + |\xi|^{m_2}\right)^{-1/2} \cdot \left[\left(|\xi|^{m_1} + |\xi|^{m_2}\right)^{1/2} \widehat{\psi}(\xi,\mu)\right] d\xi.
\end{aligned}$$

Thus, introducing the functions

$$\alpha(\lambda,\xi)=e^{i\lambda\xi}\left(|\xi|^{m_1}+|\xi|^{m_2}\right)^{-\frac{1}{2}},\quad \beta(\mu,\xi)=\left(|\xi|^{m_1}+|\xi|^{m_2}\right)^{\frac{1}{2}}\widehat{\psi}(\xi,\mu).$$

we have

$$\psi(\lambda,\mu)=\int_{\mathbb{R}}\alpha(\lambda,\xi)\beta(\mu,\xi)d\xi.$$

Moreover, the assumption together with (2.10) implies the functions α and β satisfy the condition of Theorem 2.2.4 (iii) with respect to the measure space $(\Omega, d\eta(t))=(\mathbb{R},d\xi)$. Hence, by Theorem 2.2.4, $\psi\in\mathfrak{M}$ and $\|\psi\|_{\mathfrak{M}}\leq CC_0$, where the constant C depends on m_1 and m_2 only. □

We now move on to the fundamental property of the double operator integral $T^{A,B}_{f^{[1]}}$.

Suppose that f is a differentiable function on $\mathbb{R}$. As before, we define the divided difference $f^{[1]}$ by setting

$$f^{[1]}(\lambda,\mu):=\begin{cases}\frac{f(\lambda)-f(\mu)}{\lambda-\mu}, & \text{if } \lambda\neq\mu\\ f'(\lambda), & \text{if } \lambda=\mu,\quad \lambda,\mu\in\mathbb{R}.\end{cases}\tag{2.11}$$

Theorem 2.2.9 ([BS73, Theorem 4.5] (see also [Pel16, Theorem 1.2.3])) *Let A, B be self-adjoint operators on $\mathcal{H}$ with common dense domain such that $A-B\in\mathcal{L}_1(\mathcal{H})$ (respectively, $A-B\in\mathcal{L}(\mathcal{H})$). Suppose also that the double operator integral $T^{A,B}_{f^{[1]}}$ is a bounded operator on $\mathcal{L}_1(\mathcal{H})$ (and hence on $\mathcal{L}(\mathcal{H})$). Then*

$$f(B)-f(A)=T^{A,B}_{f^{[1]}}(B-A).\tag{2.12}$$

In particular, $f(B)-f(A)\in\mathcal{L}_1(\mathcal{H})$ (respectively, $f(B)-f(A)\in\mathcal{L}(\mathcal{H})$) with

$$\|f(B)-f(A)\|_1\leq\|T^{A,B}_{f^{[1]}}\|_{\mathcal{L}(\mathcal{L}_1(\mathcal{H}))}\|B-A\|_1$$

(and

$$\|f(B)-f(A)\|\leq\|T^{A,B}_{f^{[1]}}\|_{\mathcal{L}(\mathcal{L}_1(\mathcal{H}))}\|B-A\|,$$

respectively).

Now we recall a sufficient condition on a function f, so that the double operator integral with the symbol $f^{[1]}$ is bounded on $\mathcal{L}_1(\mathcal{H})$ and on $\mathcal{L}(\mathcal{H})$. First, a function

f on $\mathbb{R}$ is of Hölder class α, $0 \leq \alpha \leq 1$, if

$$\|f\|_{\Lambda_\alpha} = \sup_{t_1,t_2} \frac{|f(t_1) - f(t_2)|}{|t_1 - t_2|^\alpha} < \infty.$$

Theorem 2.2.10 ([PS09, Theorem 4 and Corollary 2]) *Let $f : \mathbb{R} \to \mathbb{C}$. Assume that for some $0 \leq \theta < 1$ and $0 < \varepsilon \leq 1$ we have $\|f\|_{\Lambda_\theta}, \|f'\|_\infty, \|f'\|_{\Lambda_\varepsilon} < \infty$. Then the double operator integral $T^{A,B}_{f^{[1]}}$ is bounded on $\mathcal{L}_1(\mathcal{H})$ and on $\mathcal{L}(\mathcal{H})$. In particular, if $A - B \in \mathcal{L}_1(\mathcal{H})$, then $f(A) - f(B) \in \mathcal{L}_1(\mathcal{H})$ and*

$$\|f(A) - f(B)\|_1 \leq \text{const}\, \|A - B\|_1.$$

Remark 2.2.11 We note that this class of functions f, such that the double operator integral $T^{A,B}_{f^{[1]}}$ is bounded on $\mathcal{L}_1(\mathcal{H})$ is not optimal. Other possible classes are the Wiener class, the Besov class (see [Pel85, Pel90]) and there are others. In addition the boundedness of the double operator integral $T^{A,B}_{f^{[1]}}$ on $\mathcal{L}_p(\mathcal{H})$, $1 < p < \infty$, is implied by requiring that the function f is Lipschitz [PS11]. ◇

In summary, we have that for 'sufficiently nice' functions f we can express the operator $f(A) - f(B)$ in terms of an operator applied to $A - B$ and, if $A - B$ is a trace-class operator, then $f(A) - f(B)$ is trace-class too. However, for our purposes we also need to have $f(A) - f(B) \in \mathcal{L}_1(\mathcal{H})$ without the assumption that $A - B \in \mathcal{L}_1(\mathcal{H})$. We consider the construction of double operator integral suitable for this case next.

2.3 Double Operator Integrals for Resolvent Comparable Operators

In this section, we consider double operator integrals such that the bounded perturbation $A - B$ with some additional resolvent comparability condition is mapped to $f(A) - f(B)$ in the trace-class. The construction presented below is taken from [Yaf05].

We begin by introducing the resolvent comparability condition mentioned in the previous paragraph. This notion should be understood by reference to the classical 'resolvent comparability' in [Yaf92].

Definition 2.3.1 Let $m \in \mathbb{N}$. Assume that A and B are self-adjoint operators in the Hilbert space $\mathcal{H}$. We say that A, B are m-resolvent comparable (in $\mathcal{L}_1(\mathcal{H})$) if for all $a \in \mathbb{R}\backslash\{0\}$, we have

$$\left[(B - ai)^{-m} - (A - ai)^{-m}\right] \in \mathcal{L}_1(\mathcal{H}). \tag{2.13}$$

We note that by the resolvent identity it follows that if for some $z_0 \in \mathbb{C}\backslash\mathbb{R}$,

$$[(B - z_0)^{-1} - (A - z_0)^{-1}] \in \mathcal{L}_1(\mathcal{H}),$$

then

$$[(B - z)^{-1} - (A - z)^{-1}] \in \mathcal{L}_1(\mathcal{H}),$$

for all $z \in \mathbb{C}\backslash\mathbb{R}$. Thus for 1-resolvent comparable operators, it is sufficient to have the inclusion (2.13) only at one point $z \in \mathbb{C}\backslash\mathbb{R}$. However, for $m > 1$, this is no longer true as the following example demonstrates.

Example 2.3.2 ([CGL⁺16a, Example 4.2]) Suppose $\mathcal{H}$ is an infinite-dimensional Hilbert space, and let $P_j \in \mathcal{L}(\mathcal{H})$, $j \in \{1, 2\}$, be infinite-dimensional orthogonal projections satisfying

$$P_1 P_2 = 0 \quad \text{and} \quad P_1 + P_2 = 1.$$

Set

$$A = \sqrt{3}(P_1 + P_2), \quad B = \sqrt{3}(P_1 - P_2).$$

It is easy to check that

$$(A - i)^{-3} - (B - i)^{-3} = 0 \in \mathcal{L}_1(\mathcal{H}).$$

and

$$(A + 3i)^{-3} - (B + 3i)^{-3} = -\frac{1}{12\sqrt{3}} P_2 \notin \mathcal{L}_1(\mathcal{H}),$$

due to the fact that P_2 is an infinite-dimensional projection in $\mathcal{H}$.

Remark 2.3.3 For the rest of this section we assume that A and B are m-resolvent comparable operators for some *odd* $m \in \mathbb{N}$. ◇

The following construction is taken from [Yaf05]. Fix a bijection $\varphi : \mathbb{R} \to \mathbb{R}$ satisfying for some $c > 0$ and $r > 0$,

$$\varphi \in C^2(\mathbb{R}), \quad \varphi(\lambda) = \lambda^m, \ |\lambda| \geq r, \quad \varphi'(\lambda) \geq c, \ \lambda \in \mathbb{R}. \tag{2.14}$$

We choose a function $\theta \in C^2(\mathbb{R})$ such that $\theta(\lambda) = 0$ for $|\lambda| \leq r/2$, $\theta(\lambda) = 1$ for $|\lambda| \geq r$ and

$$\frac{1}{\varphi(\lambda) - i} = \theta(\lambda)\frac{1}{\lambda^m - i} + (1 - \theta(\lambda))\frac{1}{\varphi(\lambda) - i} =: g_1(\lambda) + g_2(\lambda), \quad \lambda \in \mathbb{R}.$$

Thus,

$$(\varphi(A) - i)^{-1} - (\varphi(B) - i)^{-1} = g_1(A) - g_1(B) + g_2(A) - g_2(B). \tag{2.15}$$

Next, we denote

$$\begin{aligned} G_{1,a}(\lambda, \mu) &= \frac{g_1(\lambda) - g_1(\mu)}{(\lambda - ia)^{-m} - (\mu - ia)^{-m}}, \\ G_{2,a}(\lambda, \mu) &= \frac{g_2(\lambda) - g_2(\mu)}{(\lambda - ia)^{-m} - (\mu - ia)^{-m}}, \quad \lambda, \mu \in \mathbb{R}, \end{aligned} \tag{2.16}$$

where $a \in \mathbb{R}\backslash\{0\}$.

In [Yaf05, Proposition 3.3] it is proved that there exists a (sufficiently small) $a_1 \in \mathbb{R}\backslash\{0\}$, such that the function G_{1,a_1} can be essentially represented via a function satisfying the assumption of Theorem 2.2.8. Similarly, in [Yaf05, Proposition 3.2] it is proved that there exists a (sufficiently large) $a_2 \in \mathbb{R}\backslash\{0\}$, such that the function G_{2,a_2} can be represented as a function satisfying the assumption of Theorem 2.2.8. Therefore, Theorem 2.2.8 implies that

$$g_1(A) - g_1(B) = T^{A,B}_{G_{1,a_1}}\left((A - a_1 i)^{-m} - (B - a_1 i)^{-m}\right)$$

and

$$g_2(A) - g_2(B) = T^{A,B}_{G_{2,a_2}}\left((A - a_2 i)^{-m} - (B - a_2 i)^{-m}\right)$$

with bounded (on $\mathcal{L}_1(\mathcal{H})$) double operator integrals $T^{A,B}_{G_{j,a_j}}$, $j = 1, 2$.

Thus, combining results of [Yaf05] with (2.15) one arrives at the following result.

Proposition 2.3.4 ([Yaf05]) *Suppose that A, B are m-resolvent comparable operators for some odd $m \in \mathbb{N}$ and let functions $\varphi, G_{j,a}, j = 1, 2$, be as in (2.14) and (2.16), respectively. Then there exist $a_1, a_2 \in \mathbb{R}\backslash\{0\}$ such that the double operator integrals $T^{A,B}_{G_{1,a_1}}$ and $T^{A,B}_{G_{2,a_2}}$ are bounded on $\mathcal{L}_1(\mathcal{H})$ and on $\mathcal{L}(\mathcal{H})$ and*

$$\begin{aligned} &(\varphi(A) - i)^{-1} - (\varphi(B) - i)^{-1} \\ &= T^{A,B}_{G_{1,a_1}}\left((A - a_1 i)^{-m} - (B - a_1 i)^{-m}\right) \\ &+ T^{A,B}_{G_{2,a_2}}\left((A - a_2 i)^{-m} - (B - a_2 i)^{-m}\right) \end{aligned} \tag{2.17}$$

We now define the class of functions f for which the difference $f(A) - f(B)$ is trace-class.

Definition 2.3.5 [Yaf05] Let $m \in \mathbb{N}$. Define the class of functions $\mathfrak{F}_m(\mathbb{R})$ by

$$\mathfrak{F}_m(\mathbb{R}) := \big\{ f \in C^2(\mathbb{R}) \,\big|\, f^{(\ell)} \in L_\infty(\mathbb{R}); \text{ there exists } \varepsilon > 0 \text{ and } f_0 \in \mathbb{C} \\ \text{such that } (d^\ell/d\lambda^\ell)\big[f(\lambda) - f_0\lambda^{-m}\big] \underset{|\lambda|\to\infty}{=} O\big(|\lambda|^{-\ell-m-\varepsilon}\big),\ \ell = 0, 1, 2\big\}. \tag{2.18}$$

(It is implied that f_0 is the same as $\lambda \to \pm\infty$.)

In particular, one notes that for all $m \in \mathbb{N}$,

$$S(\mathbb{R}) \subset \mathfrak{F}_m(\mathbb{R}). \tag{2.19}$$

Let $f \in \mathfrak{F}_m(\mathbb{R})$ and let φ be as before (see (2.14)). The assumptions on the functions φ and f imply that $f \circ \varphi^{-1} \in \mathfrak{F}_1(\mathbb{R})$ (see [Yaf05]). It follows from the discussion before [Yaf92, Theorem 8.7.1] that there is a continuously differentiable function h on $\mathbb{T}$, with h' satisfying the Hölder condition with exponent $\varepsilon > 0$, such that

$$f \circ \varphi^{-1} = h \circ \gamma,$$

where $\gamma(\lambda) = \frac{\lambda+i}{\lambda-i}$, $\lambda \in \mathbb{R}$, denotes the Cayley transform. Since h satisfies Hölder condition with exponent $\varepsilon > 0$, it follows from [BS66, Theorem 11] that double operator $T^{U,V}_{h^{[1]}}$ is bounded on $\mathcal{L}_p(\mathcal{H})$, $p \geq 1$ for any unitaries U, V.

Introduce the function Ψ on $\mathbb{R}^2$ by setting

$$\Psi(\lambda, \mu) = h^{[1]}(\gamma(\varphi(\lambda)), \gamma(\varphi(\mu))), \quad \lambda, \mu \in \mathbb{R}. \tag{2.20}$$

Using the change of variable in double operator integrals for unitary operators $\gamma(\varphi(A))$ and $\gamma(\varphi(B))$ (see Proposition 2.2.6 and Remark 2.2.7) we have that $T^{A,B}_\Psi = T^{\gamma(\varphi(A)),\gamma(\varphi(B))}_{h^{[1]}}$. As noted above $T^{\gamma(\varphi(A)),\gamma(\varphi(B))}_{h^{[1]}}$ is bounded on $\mathcal{L}_1(\mathcal{H})$. Hence, we infer that

$$T^{A,B}_\Psi \in \mathcal{L}(\mathcal{L}_1(\mathcal{H})). \tag{2.21}$$

Thus, we have

$$\begin{aligned} f(A) - f(B) &= h(\gamma(\varphi(A))) - h(\gamma(\varphi(B))) = T^{A,B}_\Psi(\gamma(\varphi(A)) - \gamma(\varphi(B))) \\ &= 2i T^{A,B}_\Psi\big((\varphi(A) - i)^{-1} - (\varphi(B) - i)^{-1}\big). \end{aligned}$$

Therefore, recalling (2.17)

$$\begin{aligned} f(A) - f(B) &= 2iT_\Psi^{A,B}\big((\varphi(A)-i)^{-1} - (\varphi(B)-i)^{-1}\big) \\ &= 2i(T_\Psi^{A,B} \circ T_{G_{1,a_1}}^{A,B})\big((A-a_1i)^{-m} - (B-a_1i)^{-m}\big) \\ &+ 2i(T_\Psi^{A,B} \circ T_{G_{2,a_2}}^{A,B})\big((A-a_2i)^{-m} - (B-a_2i)^{-m}\big). \end{aligned} \tag{2.22}$$

We summarise the construction in the following

Definition 2.3.6 Assume that self-adjoint operators A, B are m-resolvent comparable for some odd $m \in \mathbb{N}$ and let $f \in \mathfrak{F}_m(\mathbb{R})$. Let $a_1, a_2 \in \mathbb{R} \setminus \{0\}$ and the functions φ, G_{1,a_1}, G_{2,a_2} be as in Proposition 2.3.4. Let Ψ be defined by (2.20). Then, the double operator integrals $T_{f,a_1}^{A,B}$ and $T_{f,a_2}^{A,B}$ are defined by setting

$$T_{f,a_j}^{A,B} = 2iT_\Psi^{A,B} \circ T_{G_{j,a_j}}^{A,B}, \quad j = 1, 2. \tag{2.23}$$

Using the notation of Definition 2.3.6 and recalling Proposition 2.3.4, (2.21) and (2.22) we conclude the following result, which is proved in [Yaf05].

Proposition 2.3.7 *Assume that for some odd $m \in \mathbb{N}$ the operators A, B are m-resolvent comparable and let $f \in \mathfrak{F}_m(\mathbb{R})$. Then there exist (sufficiently small) $a_1 \in \mathbb{R} \setminus \{0\}$ and (sufficiently large) $a_2 \in \mathbb{R} \setminus \{0\}$, such that the double operator integrals $T_{f,a_1}^{A,B}$ and $T_{f,a_2}^{A,B}$, introduced in (2.23) are bounded on $\mathcal{L}_1(\mathcal{H})$ and*

$$f(A) - f(B) = \sum_{j=1,2} T_{f,a_j}^{A,B}\big((A-a_ji)^{-m} - (B-a_ji)^{-m}\big) \in \mathcal{L}_1(\mathcal{H}).$$

2.4 Continuity of Double Operator Integrals with Respect to the Operator Parameters

In this section, we discuss continuity of a double operator integral $T_\psi^{A,B}$ with respect to the operator parameters A and B.

For this we recall the question of differentiability of operator valued functions. To this end, assume that A_0 is a self-adjoint operator and let $B \in \mathcal{L}(\mathcal{H})$ be a self-adjoint perturbation. Consider the family $A(t) = A_0 + tB$ for $t \in [0, 1]$. Daletskii and Krein [DK56] established that the function $t \mapsto h(A(t))$ is differentiable (in $\mathcal{L}(\mathcal{H})$) for a restrictive class of functions h. Namely, Eq. (2.12) implies that

$$\frac{h(A(t)) - h(A(s))}{s-t} = \frac{1}{s-t} T_{h^{[1]}}^{A(t),A(s)}\big(A(t) - A(s)\big) = T_{h^{[1]}}^{A(t),A(s)}(B).$$

If h is a function, such that $T_{h^{[1]}}^{A(t),A(s)} \to T_{h^{[1]}}^{A(t),A(t)}$ pointwise on $\mathcal{L}(\mathcal{H})$ as $s \to t$, then we can pass to the limit and

$$\frac{d}{ds} h(A(s))\Big|_{s=t} = T_{h^{[1]}}^{A(t),A(t)}(B). \tag{2.24}$$

The formula (2.24) is called *Daletskii-Krein* formula and has been further extended in [BS73] and [Pel90] (where (2.24) has been proved for a function h from a Besov class). Thus, to establish operator-differentiability it suffices to show that the relevant double operator integral is continuous with respect to the operator parameter. The continuity of double operator integrals with respect to operator parameter is also crucial in establishing main results of the present lecture notes.

Results of this section rely on a special class of functions, introduced in [BS73, p. 40]. It is guaranteed that the double operator integral with a symbol from this class converges pointwise on Schatten ideals. We start with the introduction of this class of functions.

Let A_n, B_n, A, B be self-adjoint operators in the Hilbert space $\mathcal{H}$. Suppose $\varphi(\cdot, \cdot)$ admits a representation of the form

$$\varphi(\lambda, \mu) = \int_\Omega \alpha(\lambda, t)\beta(\mu, t)\, d\eta(t), \quad (\lambda, \mu) \in \mathbb{R}^2, \tag{2.25}$$

where $(\Omega, d\eta(t))$ is an auxiliary measure space and

$$C_\alpha^2 := \sup_{\lambda\in\mathbb{R}} \int_\Omega |\alpha(\lambda, t)|^2\, d\eta(t) < \infty, \quad C_\beta^2 := \sup_{\mu\in\mathbb{R}} \int_\Omega |\beta(\mu, t)|^2\, d\eta(t) < \infty.$$

We introduce

$$\varepsilon_n(v, \alpha) = \left[\int_\Omega \|\alpha(A_n, t)v - \alpha(A, t)v\|^2\, d\eta(t)\right]^{1/2},$$

$$\delta_n(v, \beta) = \left[\int_\Omega \|\beta(B_n, t) - \beta(B, t)v\|^2\, d\eta(t)\right]^{1/2}, \quad n \in \mathbb{N},\ v \in \mathcal{H}.$$

Definition 2.4.1 The class $\mathfrak{A}_r^s(E_A)$ consists of all functions φ, which admit a representation of the form (2.25), such that $\lim_{n\to\infty} \varepsilon_n(v, \alpha) = 0$ for all $v \in \mathcal{H}$. Similarly, the class $\mathfrak{A}_l^s(E_B)$ consists of all functions φ, admitting a representation (2.25), such that $\lim_{n\to\infty} \delta_n(v, \beta) = 0$ for all $v \in \mathcal{H}$.

If A_n, B_n, A, B are unitary operators on $\mathcal{H}$, the classes $\mathfrak{A}_r^s(E_A), \mathfrak{A}_l^s(E_A)$ are introduced similarly by taking the corresponding spectral measures and functions over the circle $\mathbb{T}$.

We note that if $\varphi \in \mathfrak{A}_r^s(E_A) \cap \mathfrak{A}_l^s(E_A)$ then $\lim_{n\to\infty} \varepsilon_n(v, \alpha) = 0$ and $\lim_{n\to\infty} \delta_n(v, \beta) = 0$ for all $v \in \mathcal{H}$ for two possibly different representations of φ of the form (2.25).

We also note that the definitions of the classes $\mathfrak{A}_r^s(E_A), \mathfrak{A}_l^s(E_A)$ impose certain restrictions on convergences $A_n \to A$ and $B_n \to B$ as well as on the properties of the function φ, given in (2.25). For the case of strong resolvent convergence we have the following sufficient condition.

Proposition 2.4.2 ([CGL+16a, Lemma 3.4]) *Assume that $A, A_n, n \in \mathbb{N}$, are self-adjoint operators such that $A_n \to A$ as $n \to \infty$ in the strong resolvent sense. If a function $\varphi(\cdot, \cdot)$ satisfies the condition of Theorem 2.2.8, then $\varphi \in \mathfrak{A}_r^s(E_A)$.*

Proof As shown in the proof of Theorem 2.2.8 we have

$$\varphi(\lambda, \mu) = \int_{\mathbb{R}} \alpha(\lambda, \xi)\beta(\mu, \xi)d\xi,$$

where

$$\alpha(\lambda, \xi) = e^{i\lambda\xi}\left(|\xi|^{m_1} + |\xi|^{m_2}\right)^{-\frac{1}{2}}, \quad \beta(\mu, \xi) = \left(|\xi|^{m_1} + |\xi|^{m_2}\right)^{\frac{1}{2}} \widehat{\psi}(\xi, \mu).$$

If $v \in \mathcal{H}$, then

$$\varepsilon_n(v, \alpha) = \left[\int_{\mathbb{R}} \left(|t|^{m_1} + |t|^{m_2}\right)^{-1} \left\|e^{itA_n}v - e^{itA}v\right\|^2 dt\right]^{\frac{1}{2}}, \quad n \in \mathbb{N}.$$

Fix $\delta > 0$. Since $\int_{\mathbb{R}} \left(|t|^{m_1}+|t|^{m_2}\right)^{-1} dt < \infty$ (cf., Eq. (2.10)), there exists $R > 0$ such that

$$\int_{|t|>R} \left(|t|^{m_1} + |t|^{m_2}\right)^{-1} dt < \delta.$$

Furthermore, since the family of functions $\left\{e^{i\lambda t}\right\}_{t\in[-R,R]}$ is uniformly continuous, [RS80, Theorem VIII.21] and the comment following its proof guarantees for each $v \in \mathcal{H}$,

$$\lim_{n\to 0} \left\|e^{itA_n}v - e^{itA}v\right\| = 0,$$

uniformly in $t \in [-R, R]$. Therefore, for each $v \in \mathcal{H}$, there exists $N \in \mathbb{N}$ such that

$$\left\|e^{itA_n}v - e^{itA}v\right\| < \delta, \quad n \geq N, \ t \in [-R, R].$$

Hence, for every $v \in \mathcal{H}$,

$$\lim_{n\to\infty} \varepsilon_n(v,\alpha) \le \lim_{n\to\infty} \left[\int_{|t|\le R} \|e^{itA_n}v - e^{itA}v\|^2\, dt \right]^{\frac{1}{2}} + \lim_{n\to\infty} \left[\int_{|t|>R} \|e^{itA_n}v - e^{itA}v\|^2\, dt \right]^{\frac{1}{2}} \le 2\delta \|v\|.$$

Since $\delta > 0$ is arbitrary, one concludes

$$\lim_{n\to\infty} \varepsilon_n(v,\alpha) = 0, \quad v \in \mathcal{H}.$$

□

The crucial property of the classes $\mathfrak{A}_r^s(E_A)$ and $\mathfrak{A}_l^s(E_B)$ is the following convergence result.

Proposition 2.4.3 ([BS73, Proposition 5.6]) *Let* $\psi \in \mathfrak{A}_r^s(E_A) \cap \mathfrak{A}_l^s(E_B)$. *Then* $T_\psi^{A_n,B_m} \to T_\psi^{A,B}$ *pointwise on* $\mathcal{L}_1(\mathcal{H})$.

Proof Without loss of generality ψ is real-valued. Since $\psi \in \mathfrak{A}_r^s(E_A) \cap \mathfrak{A}_l^s(E_B)$ we can write representations of ψ in the form

$$\psi(\lambda,\mu) = \int_\Omega \alpha(\lambda,t)\tilde{\beta}(\mu,t)d\eta(t) = \int_\Omega \tilde{\alpha}(\lambda,t)\beta(\mu,t)d\eta(t),$$

such that $\lim_{n\to\infty} \varepsilon_n(v,\alpha) = \lim_{n\to\infty} \delta_n(v,\beta) = 0$ for all $v \in \mathcal{H}$.

Since $\mathcal{L}_1(\mathcal{H})$ is a separable ideal and

$$\sup_{n,m\in\mathbb{N}} \|T_\psi^{A_n,B_m}\|_{\mathcal{L}(\mathcal{L}_1(\mathcal{H}))} \le \|\psi\|_{\mathfrak{M}} < \infty,$$

it is sufficient to show that $T_\psi^{A_n,B_m}(X) \to T_\psi^{A,B}(X)$ in $\mathcal{L}_1(\mathcal{H})$ for any rank-one operator X.

Let $T \in \mathcal{L}(\mathcal{H})$ and $X = \langle \cdot, e\rangle g$, $e, g \in \mathcal{H}$. By (2.6) we have

$$\begin{aligned}
&\operatorname{tr}\Big(T \cdot (T_\psi^{A_n,B} - T_\psi^{A,B})(X)\Big) \overset{(2.6)}{=} \operatorname{tr}\Big((T_\psi^{A_n,B} - T_\psi^{A,B})(T) \cdot X\Big)\\
&= \langle (T_\psi^{A_n,B} - T_\psi^{A,B})(T)g, e\rangle = \langle \int_\Omega \big(\alpha(A_n,t) - \alpha(A,t)\big)T\tilde{\beta}(B,t)g d\eta(t), e\rangle\\
&= \int_\Omega \langle T\tilde{\beta}(B,t)g, \big(\alpha(A_n,t) - \alpha(A,t)\big)e\rangle d\eta(t)
\end{aligned}$$

Therefore,

$$\begin{aligned}
\|T_\psi^{A_n,B}(X) - T_\psi^{A,B}(X)\|_1 &= \sup_{\|T\|\le 1} \left|\operatorname{tr}\left(T\cdot (T_\psi^{A_n,B} - T_\psi^{A,B})(X)\right)\right| \\
&\le \sup_{\|T\|\le 1} \int_\Omega \|T\|\|\tilde\beta(B,t)\|\|g\| \left\|(\alpha(A_n,t) - \alpha(A,t))e\right\| d\eta(t) \\
&\le \left(\int_\Omega \|\tilde\beta(B,t)\|^2 d\eta(t)\right)^{1/2} \left(\int_\Omega \left\|(\alpha(A_n,t) - \alpha(A,t))e\right\|^2 d\eta(t)\right)^{1/2} \|g\| \\
&\le C_{\tilde\beta}\varepsilon_n(e,\alpha)\|g\|.
\end{aligned}$$

Since $\varepsilon_n(e,\alpha)\to 0$, it follows that

$$\|T_\psi^{A_n,B}(X) - T_\psi^{A,B}(X)\|_1 \to 0, \quad n\to\infty. \tag{2.26}$$

Similarly,

$$\begin{aligned}
&\|T_\psi^{A_n,B_m}(X) - T_\psi^{A_n,B}(X)\|_1 \\
&= \sup_{\|T\|\le 1} \left|\operatorname{tr}\left((T_\psi^{A_n,B_m} - T_\psi^{A_n,B})(T)\cdot X\right)\right| \\
&= \sup_{\|T\|\le 1} \left|\langle \int_\Omega \tilde\alpha(A_n,t)T\big(\beta(B_m,t) - \beta(B,t)\big)g\, d\eta(t), e\rangle\right| \\
&\le \sup_{\|T\|\le 1} \int_\Omega \|\tilde\alpha(A_n,t)\|\|T\|\left\|\Big(\beta(B_m,t) - \beta(B,t)\Big)g\right\|g\|\|e\|\, d\eta(t) \\
&\le C_{\tilde\alpha}\delta_m(g,\beta)\|e\| \to 0.
\end{aligned}$$

Combining the latter convergence with (2.26) we conclude that

$$\|T_\psi^{A_n,B_m}(X) - T_\psi^{A,B}(X)\|_1 \le \|T_\psi^{A_n,B}(X) - T_\psi^{A,B}(X)\|_1 + \|T_\psi^{A_n,B_m}(X) - T_\psi^{A_n,B}(X)\|_1 \to 0,$$

as required. □

For the purpose of these notes we recall a result about continuity of the double operator integral for a very restrictive class of functions f, which is sufficient to establish the principal trace formula (1.2).

Assume now that $f\in C_b^2(\mathbb{R})$ be such that $f'\in L_p(\mathbb{R})$ for some $p\ge 1$ and that $\{A_i\}_{i\in I}, \{B_i\}_{i\in I}$ and A, B are self-adjoint operators on $\mathcal{H}$, such that $A_i\to A$, $B_i\to B$ in the strong resolvent sense. Then, as follows from [BS73, Proposition 7.8, Theorem 5.7], the divided difference $f^{[1]}$ belongs to the class $\mathfrak{A}_r^s(E_A)$. Since $f^{[1]}(\lambda,\mu) = f^{[1]}(\mu,\lambda)$ for all $\lambda,\mu\in\mathbb{R}$, it follows that $f^{[1]}\in\mathfrak{A}_l^s(E_B)$. Hence, Proposition 2.4.3 implies the following sufficient result for convergence of double operator integrals.

Proposition 2.4.4 *Let $f \in C_b^2(\mathbb{R})$ be such that $f' \in L_p(\mathbb{R})$ for some $p \geq 1$ and let $\{A_i\}_{i\in I}, \{B_i\}_{i\in I}$ and A, B be self-adjoint operators on $\mathcal{H}$, such that $A_i \to A$, $B_i \to B$ in the strong resolvent sense. Then the double operator integrals $T_{f^{[1]}}^{A_i,B_j}$ converge to $T_{f^{[1]}}^{A,B}$ pointwise on $\mathcal{L}_1(\mathcal{H})$.*

A similar result holds for double operator integrals constructed over spectral measure of unitary operators.

Proposition 2.4.5 ([BS73, Proposition 7.5, Theorem 5.9]) *Let h be such that h' satisfies the Hölder condition with exponent $\varepsilon > 0$ and let $\{U_n\}_{n\in\mathbb{N}}, \{V_n\}_{n\in\mathbb{N}}, U, V$, be unitary operators such that $U_n \to U$ and $V_n \to V$ in the strong operator topology. Then $T_{h^{[1]}}^{U_n,V_m} \to T_{h^{[1]}}^{U,V}$ pointwise on $\mathcal{L}_1(\mathcal{H})$ as $n, m \to \infty$.*

As a corollary of Proposition 2.4.4 we prove a formula, which is fundamental to deriving the right-hand side of the principal trace formula. The following proposition is an easy corollary of the Daletski-Krein formula. We refer also to [Sim98]. For completeness we present the full proof.

Proposition 2.4.6 *Let A_0 be a self-adjoint operator acting in a separable Hilbert space $\mathcal{H}$, and let $f \in C_b^2(\mathbb{R})$ be such that $f' \in L_p(\mathbb{R}) \cap \mathrm{Lip}(\mathbb{R})$ for some $p \geq 1$. Assume, in addition, that $\{B_s\}_{s\in[0,1]} \subset \mathcal{L}_1(\mathcal{H})$ is ($\mathcal{L}_1(\mathcal{H})$-)differentiable function. Then, letting $A_s = A_0 + B_s$, $s \in [0, 1]$, we have that*

$$\mathrm{tr}(f(A_1) - f(A_1)) = \int_0^1 \mathrm{tr}\left(f'(A_s)B_s'\right)ds.$$

Proof By Theorem 2.2.10, the double operator integral $T_{f^{[1]}}^{A_s,A_t}$ is bounded on $\mathcal{L}_1(\mathcal{H})$ for any $s, t \in [0, 1]$-function.

$$s \mapsto \mathrm{tr}(f(A_s) - f(A_0))$$

is a $C^1[0, 1]$-function.

Let $s, t \in [0, 1]$. We have

$$\begin{aligned}
|\,\mathrm{tr}(f(A_s) - f(A_0)) - \mathrm{tr}(f(A_t) - f(A_0))| &= |\,\mathrm{tr}(f(A_s) - f(A_t))| \\
&= \left|\mathrm{tr}(T_{f^{[1]}}^{A_s,A_t}(A_s - A_t))\right| \leq \|T_{f^{[1]}}^{A_s,A_t}\|_{\mathcal{L}(\mathcal{L}_1(\mathcal{H}))}\|A_s - A_t\|_1 \\
&\leq \mathrm{const}\,\|B_s - B_t\|_1,
\end{aligned}$$

where the constant on the right-hand side does not depend on s and t. Therefore, we infer that the function $s \mapsto \mathrm{tr}(f(A_s) - f(A_0))$ is continuous.

To prove that this function is continuously differentiable, we write

$$\begin{aligned}
&\frac{1}{s-t}\cdot\big(\mathrm{tr}(f(A_s)-f(A_0))-\mathrm{tr}(f(A_t)-f(A_0))\big)\\
&\quad=\frac{1}{s-t}\,\mathrm{tr}(f(A_s)-f(A_t))=\mathrm{tr}\left(T^{A_s,A_t}_{f^{[1]}}(\frac{B_s-B_t}{s-t})\right)\\
&\quad=\mathrm{tr}\left(T^{A_s,A_t}_{f^{[1]}}(\frac{B_s-B_t}{s-t}-B_t')\right)+\mathrm{tr}\left(T^{A_s,A_t}_{f^{[1]}}(B_t')\right).
\end{aligned}$$

Since the function f satisfies the assumptions in Proposition 2.4.4, we infer that

$$\lim_{s\to t}\|T^{A_s,A_t}_{f^{[1]}}(B_t')-T^{A_t,A_t}_{f^{[1]}}(B_t')\|_1=0.$$

Moreover,

$$\left|\mathrm{tr}\left(T^{A_s,A_t}_{f^{[1]}}(\frac{B_s-B_t}{s-t}-B_t')\right)\right|\le\left\|T^{A_s,A_t}_{f^{[1]}}(\frac{B_s-B_t}{s-t}-B_t')\right\|_1\le\mathrm{const}\left\|\frac{B_s-B_t}{s-t}-B_t'\right\|_1\to 0,$$

where the constant on the right-hand side does not depend on s and t. Hence,

$$\frac{d}{ds}\mathrm{tr}(f(A_s)-f(A_0))|_{s=t}=\mathrm{tr}\left(T^{A_t,A_t}_{f^{[1]}}(B_t')\right)=\mathrm{tr}\left(T^{A_t,A_t}_{f^{[1]}}(1)\cdot B_t'\right),$$

where the latter equality follows from the definition of the double operator integral $T^{A_t,A_t}_{f^{[1]}}$ on $\mathcal{L}(\mathcal{H})$ (see (2.6)). Moreover, we infer that the function $s\mapsto\mathrm{tr}(f(A_s)-f(A_0))$ is a $C^1[0,1]$-function.

Hence, by the fundamental theorem of calculus and Proposition 2.2.5 we have

$$\begin{aligned}
\mathrm{tr}(f(A_1)-f(A_0))&=\int_0^1\frac{d}{ds}\mathrm{tr}(f(A_s)-f(A_0))ds\\
&=\int_0^1\mathrm{tr}\left(T^{A_s,A_s}_{f^{[1]}}(1)\cdot B_s'\right)ds\\
&=\int_0^1\mathrm{tr}\left(f'(A_s)\cdot B_s'\right)ds,
\end{aligned}$$

as required. □

Next, we recall the continuity of the double operator integrals, introduced in Definition 2.3.6, with respect to the operator parameter, which was proved in [CGL+16a]. To this end we need the following assumption.

Hypothesis 2.4.7 *Assume that A, A_n, B, B_n, $n \in \mathbb{N}$, are self-adjoint operators such that $A_n \to A$ and $B_n \to B$ as $n \to \infty$ in the strong resolvent sense. In addition we assume that for some $m \in \mathbb{N}$, m odd, and every $a \in \mathbb{R}\backslash\{0\}$,*

$$\big[(A+ia)^{-m}-(B+ia)^{-m}\big],\quad \big[(A_n+ia)^{-m}-(B_n+ia)^{-m}\big]\in\mathcal{L}_1(\mathcal{H}),$$

and

$$\lim_{n\to\infty}\big\|\big[(A_n+ia)^{-m}-(B_n+ia)^{-m}\big]-\big[(A+ia)^{-m}-(B+ia)^{-m}\big]\big\|_1=0.$$

With this hypothesis at hand, the following theorem is the main result of the main result of [CGL+16a, Section 3].

Theorem 2.4.8 ([CGL+16a, Section 3]) *Assume Hypothesis 2.4.7. Let $f \in \mathfrak{F}_m(\mathbb{R})$ and let the double operator integrals $T^{A,B}_{f,a_j}$ and T^{A_n,B_n}_{f,a_j} be as in Definition 2.3.6 (with A, B replaced by A_n, B_n respectively for T^{A_n,B_n}_{f,a_j}). Then*

$$T^{A_n,B_n}_{f,a_j}\to T^{A,B}_{f,a_j},\quad j=1,2,$$

pointwise on $\mathcal{L}_1(\mathcal{H})$.

We briefly sketch the proof of Theorem 2.4.8 for $j = 1$. The proof for $j = 2$ is identical.

By definition (see Definition 2.3.6) we have that

$$T^{A,B}_{f,a_1}=2iT^{A,B}_{\Psi}\circ T^{A,B}_{G_{1,a_1}},$$

and

$$T^{A_n,B_n}_{f,a_1}=2iT^{A_n,B_n}_{\Psi}\circ T^{A_n,B_n}_{G_{1,a_1}}.$$

As shown in [Yaf05, Proposition 3.3] the function G_{1,a_1} can be represented via a function satisfying the assumption of Theorem 2.2.8. By Proposition 2.4.2 this implies that $G_{1,a_1} \in \mathfrak{A}^s_r(E_A)$. Since $G_{1,a_1}(\lambda,\mu) = G_{1,a_1}(\mu,\lambda)$ for all $\lambda,\mu \in \mathbb{R}$, it follows that $G_{1,a_1} \in \mathfrak{A}^s_l(E_B)$. Furthermore, Proposition 2.4.5 implies that $\Psi \in \mathfrak{A}^s_r(E_A)\cap\mathfrak{A}^s_l(E_B)$. Hence, Proposition 2.4.3 implies that

$$T^{A_n,B_n}_{G_{1,a_1}}\to T^{A,B}_{G_{1,a_1}},\quad T^{A_n,B_n}_{\Psi}\to T^{A,B}_{\Psi}$$

pointwise on $\mathcal{L}_1(\mathcal{H})$.

Hence, for any $X \in \mathcal{L}_1(\mathcal{H})$ we have

$$\begin{aligned}
&\|T_{f,a_1}^{A_n,B_n}(X) - T_{f,a_1}^{A,B}(X)\|_1 \\
&\quad = 2\left\|T_\Psi^{A_n,B_n}\left(T_{G_{1,a_1}}^{A_n,B_n}(X) - T_{G_{1,a_1}}^{A,B}(X)\right)\right\|_1 + 2\left\|\left(T_\Psi^{A_n,B_n} - T_\Psi^{A,B}\right)(T_{G_{1,a_1}}^{A,B}(X))\|_1 \\
&\quad \leq 2\|\Psi\|_{\mathfrak{M}}\left\|T_{G_{1,a_1}}^{A_n,B_n}(X) - T_{G_{1,a_1}}^{A,B}(X)\right\|_1 + 2\left\|\left(T_\Psi^{A_n,B_n} - T_\Psi^{A,B}\right)(T_{G_{1,a_1}}^{A,B}(X))\right\|_1.
\end{aligned}$$

The pointwise convergence $T_{G_{1,a_1}}^{A_n,B_n} \to T_{G_{1,a_1}}^{A,B}$ guarantees that the first term vanishes. The pointwise convergence $T_\Psi^{A_n,B_n} \to T_\Psi^{A,B}$ together with the fact that $T_{G_{1,a_1}}^{A,B}(X) \in \mathcal{L}_1(\mathcal{H})$ implies that the second term vanishes too. Thus, $T_{f,a_1}^{A_n,B_n} \to T_{f,a_1}^{A,B}$ pointwise on $\mathcal{L}_1(\mathcal{H})$, as stated.

The following result is an immediate corollary of Theorem 2.4.8.

Theorem 2.4.9 ([CGL+16a, Theorem 3.7]) *Assume Hypothesis 2.4.7. Then for any function $f \in \mathfrak{F}_m(\mathbb{R})$,*

$$\lim_{n\to\infty} \left\|[f(A_n) - f(B_n)] - [f(A) - f(B)]\right\|_1 = 0.$$

Proof By Proposition 2.3.7 we have that

$$f(A) - f(B) = \sum_{j=1,2} T_{f,a_j}^{A,B}\left((A - a_j i)^{-m} - (B - a_j i)^{-m}\right),$$

$$f(A_n) - f(B_n) = \sum_{j=1,2} T_{f,a_j}^{A_n,B_n}\left((A_n - a_j i)^{-m} - (B_n - a_j i)^{-m}\right).$$

Hence, we can write

$$\begin{aligned}
&[f(A_n) - f(B_n)] - [f(A) - f(B)] \\
&= \sum_{j=1,2}\left(T_{f,a_j}^{A_n,B_n}\left((A_n - a_j i)^{-m} - (B_n - a_j i)^{-m}\right) - T_{f,a_j}^{A,B}\left((A - a_j i)^{-m} - (B - a_j i)^{-m}\right)\right) \\
&= \sum_{j=1,2} T_{f,a_j}^{A_n,B_n}\left(\left((A_n - a_j i)^{-m} - (B_n - a_j i)^{-m}\right) - \left((A - a_j i)^{-m} - (B - a_j i)^{-m}\right)\right) \\
&\quad + \sum_{j=1,2}\left(T_{f,a_j}^{A_n,B_n} - T_{f,a_j}^{A,B}\right)\left((A - a_j i)^{-m} - (B - a_j i)^{-m}\right).
\end{aligned}$$

By assumption of Hypothesis 2.4.7 we have that $\left(A_n - a_j i)^{-m} - (B_n - a_j i)^{-m}\right) \to \left((A - a_j i)^{-m} - (B - a_j i)^{-m}\right)$ in $\mathcal{L}_1(\mathcal{H})$. Since, in addition, the sequence $\{\|T_{f,a_j}^{A_n,B_n}\|_{\mathcal{L}(\mathcal{L}_1(\mathcal{H}))}\}_{n\in\mathbb{N}}$ is uniformly bounded, we conclude that the first term on the right hand side above converges to 0 in $\mathcal{L}_1(\mathcal{H})$. The second term converges to 0 in $\mathcal{L}_1(\mathcal{H})$ since $\left((A - a_j i)^{-m} - (B - a_j i)^{-m}\right) \in \mathcal{L}_1(\mathcal{H})$ and by Theorem 2.4.8 we have that $T_{f,a_j}^{A_n,B_n} \to T_{f,a_j}^{A,B}$ pointwise on $\mathcal{L}_1(\mathcal{H})$. This completes the proof. □

Chapter 3
The Model Operator and Its Approximants

In the present chapter we introduce the set-up for the rest of these notes as well as explain the approximation scheme we employ in our approach. The idea is to introduce a spectral 'cut-off' by the characteristic function of the interval $[-n, n]$ for a positive integer n. We use the subscript n on the operators introduced in Chap. 3 to indicate these spectrally cut-off or 'reduced' versions.

In Sect. 3.2 we introduce the path $\{B(t)\}_{t\in\mathbb{R}}$ of perturbations of the fixed self-adjoint operator A_- and state our initial hypothesis, Hypothesis 3.2.1. Introducing the path $\{A(t)\}_{t\in\mathbb{R}} = \{A_- + B(t)\}_{t\in\mathbb{R}}$ we can define the model operator

$$\boldsymbol{D}_{\boldsymbol{A}} = \frac{d}{dt} + \boldsymbol{A}.$$

In particular, Hypothesis 3.2.1 implies that the path $\{B(t)\}_{t\in\mathbb{R}}$, as well as its norm limit B_+ as $t \to \infty$, are p-relative trace-class perturbation of A_-. Hence, the asymptote $A_+ = A_- + B_+$ of $\{A(t)\}_{t\in\mathbb{R}}$ as $t \to +\infty$ is a perturbations of A_- by a p-relative trace-class operator, which allows us to apply the methods of Sect. 3.1 to the operators A_+ and A_-. This gives us approximation results whenever we deal with the path $\{A(t)\}_{t\in\mathbb{R}}$.

Furthermore, introducing the reduced path $\{B_n(t)\}_{t\in\mathbb{R}}$ using the spectral cut-off, we can apply results of [Pus08, GLM+11, CGP+17] to the reduced counterparts $A_{+,n}$, A_- of A_+, A_-. Together with the approximation result mentioned above this provides all technical results for the path $\{A(t)\}_{t\in\mathbb{R}}$, which are necessary in the proof of our main results.

However, to work with the operator $\boldsymbol{D}_{\boldsymbol{A}}$ we need additional technical results. In general, the operator $\boldsymbol{D}_{\boldsymbol{A}}^*\boldsymbol{D}_{\boldsymbol{A}} - \boldsymbol{D}_{\boldsymbol{A}}\boldsymbol{D}_{\boldsymbol{A}}^*$ is not a q-relative trace-class perturbation of either $\boldsymbol{D}_{\boldsymbol{A}}^*\boldsymbol{D}_{\boldsymbol{A}}$ or $\boldsymbol{D}_{\boldsymbol{A}}\boldsymbol{D}_{\boldsymbol{A}}^*$ for any $q \in \mathbb{N}$. Hence, to be able to obtain approximation technique for the operator $\boldsymbol{D}_{\boldsymbol{A}}$ and its reduced counterpart $\boldsymbol{D}_{\boldsymbol{A}_n}$ we establish some additional estimates. We introduce our main assumptions, summarised as Hypothesis 3.2.5, which apply throughout the rest of these notes. We also present a simplified version of Hypothesis 3.2.5 for the special type of path, discussed in the

A. Carey, G. Levitina, *Index Theory Beyond the Fredholm Case*, Lecture Notes in Mathematics 2323, https://doi.org/10.1007/978-3-031-19436-8_3

introduction, $\{B(t)\}_{t\in\mathbb{R}}$ consisting of operators of the form $B(t) = \theta(t)B_+$, where B_+ is a fixed p-relative trace-class perturbation of A_- and θ is a certain smooth function.

3.1 The Class of p-Relative Trace-Class Perturbations

In this section we introduce the class of perturbations needed for the rest of these notes.

Definition 3.1.1 Let A_0 be a self-adjoint operator on a separable Hilbert space $\mathcal{H}$ and let $p \in \mathbb{N}\cup\{0\}$. A bounded self-adjoint operator $B \in \mathcal{L}(\mathcal{H})$ is called a p-relative trace-class perturbation (with respect to A_0) if

$$B(A_0 + i)^{-p-1} \in \mathcal{L}_1(\mathcal{H}). \tag{3.1}$$

In what follows, we choose the smallest p, such that (3.1) holds.

It is clear that one can take the resolvent parameter for the operator A_0 to be any point $z \in \mathbb{C}\setminus\mathbb{R}$ for the definition of p-relative trace-class perturbations. In addition, since B is self-adjoint, inclusion (3.1) is equivalent to the inclusion $(A_0+i)^{-p-1}B \in \mathcal{L}_1(\mathcal{H})$.

Example 3.1.2 Let $\mathcal{D}$ be the (massive or massless) Dirac operator on $\mathbb{R}^d$, $d \in \mathbb{N}$, and let $V = \{\varphi_{ij}\}_{i,j=1}^{n(d)}$ be the perturbation of $\mathcal{D}$ such that $|\varphi_{ij}(x)| \leq \mathrm{const}\langle x\rangle^{-\rho}$ for some $\rho > d$ (see Sect. 7.1 for precise definitions). As we will show in Proposition 7.1.2, V is a d-relative trace-class perturbation of $\mathcal{D}$.

Assume that B is a p-relative trace-class perturbation of A_0. By [GLST15, Theorem 3.2] we have that

$$(A_0 + i)^{-k}B(A_0 + i)^{-l} \in \mathcal{L}_{\frac{p+1}{j}}(\mathcal{H}), \tag{3.2}$$

and

$$\|(A_0+i)^{-k}B(A_0+i)^{-l}\|_{\frac{p+1}{j}} \leq \|B(A_0+i)^{-j}\|_{\frac{p+1}{j}} \leq \|B\|^{\frac{j}{p+1}} \cdot \|B(A_0+i)^{-p+1}\|_1^{\frac{p+1-j}{p+1}}. \tag{3.3}$$

for any $j = 1, \ldots, p+1$ and any $k, l \in \mathbb{N}\cup\{0\}$ such that $k + l = j$.

Inclusion (3.2) (for $k = 0$ and $l = 1$) combined with the stability of the essential spectrum implies, in particular, that

$$\sigma_{\mathrm{ess}}(A_0 + B) = \sigma_{\mathrm{ess}}(A_0). \tag{3.4}$$

The main purpose of this section is to show that for any p-relative trace-class perturbation B of a self-adjoint operator A_0, the pair A_0 and $A := A_0 + B$ is

p-resolvent comparable (see Definition 2.3.1), that is $(A-ai)^{-p}-(A_0-ai)^{-p}, a \in \mathbb{R}$, is a trace-class operator and that the operator $(A - ai)^{-p} - (A_0 - ai)^{-p}$ can be approximated by the $(A_n - ai)^{-p} - (A_0 - ai)^{-p}$ in the trace-class norm for an appropriate approximation A_n of A. This result is the foundation of our approximation scheme for the proof of the principal trace formula.

Firstly, we recall a result which is used repeatedly throughout these notes. This result can be found in [Sim05] and [GLM+11, Lemma 3.4].

Lemma 3.1.3 *Let $p \in [1, \infty)$ and assume that $R, R_n, T, T_n \in \mathcal{L}(\mathcal{H})$, $n \in \mathbb{N}$, satisfy*

$$\operatorname*{s-lim}_{n\to\infty} R_n = R, \ \operatorname*{s-lim}_{n\to\infty} T_n = T$$

and that $S, S_n \in \mathcal{L}_p(\mathcal{H})$, $n \in \mathbb{N}$, satisfy $\lim_{n\to\infty} \|S_n - S\|_p = 0$. Then $\lim_{n\to\infty} \|R_n S_n T_n^* - RST^*\|_p = 0$.

We now introduce special spectral 'cut-off' approximants B_n, $n \in \mathbb{N}$, for a p-relative trace-class perturbation B.

Definition 3.1.4 Let A_0 be a self-adjoint operator in $\mathcal{H}$ and let B be a p-relative trace-class perturbation of A_0. For every $n \in \mathbb{N}$ we introduce

$$P_n = \chi_{[-n,n]}(A_0) \tag{3.5}$$

and

$$B_n := P_n B P_n.$$

Throughout this section we will assume that P_n and B_n, $n \in \mathbb{N}$, are as in the above definition.

It follows from spectral theory that

$$\operatorname*{s-lim}_{n\to\infty} P_n = 1.$$

We also note that (3.1) together with the definition of the projections P_n implies that

$$B_n = P_n B P_n \in \mathcal{L}_1(\mathcal{H}). \tag{3.6}$$

Remark 3.1.5 The precise form of the cut-offs P_n is of course immaterial. We just need several facts: that $\operatorname{s-lim}_{n\to\infty} P_n = 1$, $\sup_{n\in\mathbb{N}} \|P_n\| < \infty$ and that $P_n B P_n \in \mathcal{L}_1(\mathcal{H})$. ◇

The following lemma gathers some simple properties of the approximants $A_0 + B_n$, $n \in \mathbb{N}$.

Lemma 3.1.6 *Let B, B_n and P_n be as in Definition 3.1.4. We have that*

1. *$A_0 + B_n \to A_0 + B$ in the strong resolvent sense as $n \to \infty$;*
2. *Let $j \in \mathbb{N}$. For any $k, l \in \mathbb{N}$ such that $k + l \geq j$ and $z \in \mathbb{C} \setminus \mathbb{R}$, we have*

$$\lim_{n\to\infty} \left\| (A_0 - z)^{-k} B_n (A_0 - z)^{-l} - (A_0 - z)^{-k} B (A_0 - z)^{-l} \right\|_{\frac{p+1}{j}} = 0.$$

Proof

(i) Since $P_n \to 1$ in the strong operator topology, it follows that $A_0+B_n \to A_0+B$ pointwise on the common core $\operatorname{dom}(A_0)$. Hence, the assertion follows from [RS80, Theorem VIII.25 (a)].

(ii) Since $k + l \geq j$, (3.2) implies that

$$(A_0 - z)^{-k} B (A_0 - z)^{-l} \in \mathcal{L}_{\frac{p+1}{j}}(\mathcal{H}).$$

Therefore, since

$$(A_0 - z)^{-k} B_n (A_0 - z)^{-l} = P_n (A_0 - z)^{-k} B (A_0 - z)^{-l} P_n,$$

and $P_n \to 1$ in the strong operator topology, it follows from Lemma 3.1.3 that

$$(A_0 - z)^{-k} B_n (A_0 - z)^{-l} \to (A_0 - z)^{-k} B (A_0 - z)^{-l}$$

in $\mathcal{L}_{\frac{p+1}{j}}(\mathcal{H})$ as $n \to \infty$.

□

The following theorem establishes, in particular, that the p-relative trace-class assumption implies that $A_0 + B$ and A_0 are p-resolvent comparable and the difference $(A_0 + B - z)^{-p} - (A_0 - z)^{-p}$ can be approximated (in the trace-class norm) by $(A_0 + B_n - z)^{-p} - (A_0 - z)^{-p}$. This theorem is the first key result used in the proof of the principal trace formula. This result guarantees that Hypothesis 2.4.7 is satisfied for the operators A_0, $A_0 + B$ and $A_0 + B_n$, and this fact allows us to use Theorem 2.4.8 and Theorem 2.4.9.

Theorem 3.1.7 *Assume that B is a p-relative trace-class perturbation of A_0. Set $P_n = \chi_{[-n,n]}(A_0)$ and $B_n := P_n B P_n$ and let $z \in \mathbb{C} \setminus \mathbb{R}$. For any $j = 1, \ldots, p$ we have that*

$$(A_0 + B - z)^{-j} - (A_0 - z)^{-j} \in \mathcal{L}_{\frac{p+1}{j+1}}(\mathcal{H})$$

and

$$\lim_{n\to\infty}\left\|\left[(A_0+B_n-z)^{-j}-(A_0-z)^{-j}\right]-\left[(A_0+B-z)^{-j}-(A_0-z)^{-j}\right]\right\|_{\frac{p+1}{j+1}}=0.$$

For any $j>p$ the assertion holds in the trace-class ideal $\mathcal{L}_1(\mathcal{H})$.

Proof We prove the assertion by induction on j.

Let firstly $j=1$. Applying the resolvent identity twice we have

$$\begin{aligned}(A_0+B-z)^{-1}-(A_0-z)^{-1}&=-(A_0+B-z)^{-1}B(A_0-z)^{-1}\\&=-(A_0-z)^{-1}B(A_0-z)^{-1}+\big((A_0+B-z)^{-1}-(A_0-z)^{-1}\big)B(A_0-z)^{-1}\\&=-(A_0-z)^{-1}B(A_0-z)^{-1}+(A_0+B-z)^{-1}B(A_0-z)^{-1}B(A_0-z)^{-1}\end{aligned}$$

By (3.2) (with $k=l=1$) we have that the first term belongs to $\mathcal{L}_{\frac{p+1}{2}}(\mathcal{H})$. Combining (3.2) (for $k=0, l=1$) with the Hölder inequality we infer that the second term is also in $\mathcal{L}_{\frac{p+1}{2}}(\mathcal{H})$.

To prove the convergence, using similar argument, together with the fact that P_n commutes with A_0 we write

$$\begin{aligned}&(A_0+B_n-z)^{-1}-(A_0-z)^{-1}\\&\quad=-P_n\,(A_0-z)^{-1}\,B\,(A_0-z)^{-1}\,P_n\\&\qquad+(A_0+B_n-z)^{-1}\,P_nB\,(A_0-z)^{-1}\,P_nB\,(A_0-z)^{-1}\,P_n.\end{aligned}$$

Since $P_n\to 1$ in the strong operator topology and $A_0+B_n\to A_0+B$ in the strong resolvent sense (see Lemma 3.1.6), Lemma 3.1.3 together with inclusion (3.2) implies the required convergence of each term.

Next, assume that the assumption of the theorem holds for all $k\leq j-1$. For $k=j$ using the resolvent identity we can write

$$\begin{aligned}&(A_0+B-z)^{-j}-(A_0-z)^{-j}\\&\quad=\sum_{k_0+k_1=j-1}(A_0+B-z)^{-k_0}\Big((A_0+B-z)^{-1}-(A_0-z)^{-1}\Big)(A_0-z)^{-k_1}\\&\quad=-\sum_{k=1}^{j}(A_0+B-z)^{-k}B(A_0-z)^{-j-1+k}\end{aligned}\tag{3.7}$$

and similarly

$$(A_0+B_n-z)^{-j}-(A_0-z)^{-j}=-\sum_{k=1}^{j}(A_0+B_n-z)^{-k}P_nB(A_0-z)^{-j-1+k}P_n.\tag{3.8}$$

For $k = 1, \dots, j-1$ we write

$$\begin{aligned}(A_0 + B - z)^{-k} \quad & B(A_0 - z)^{-j-1+k} \\ &= \Big((A_0 + B - z)^{-k} - (A_0 - z)^{-k}\Big) B(A_0 - z)^{-j-1+k} \\ &+ (A_0 - z)^{-k} B(A_0 - z)^{-j-1+k}\end{aligned}$$

By (3.2) we have that

$$(A_0 - z)^{-k} B(A_0 - z)^{-j-1+k} \in \mathcal{L}_{\frac{p+1}{j+1}}(\mathcal{H}).$$

In addition, by the assumption of induction we have that

$$\Big((A_0 + B - z)^{-k} - (A_0 - z)^{-k}\Big) \in \mathcal{L}_{\frac{p+1}{k+1}}(\mathcal{H})$$

and (3.2) we have $B(A_0 - z)^{-j-1+k} \in \mathcal{L}_{\frac{p+1}{j+1-k}}(\mathcal{H})$. Hence, by Hölder inequality we obtain that

$$\Big((A_0 + B - z)^{-k} - (A_0 - z)^{-k}\Big) B(A_0 - z)^{-j-1+k} \in \mathcal{L}_{\frac{p+1}{j+2}}(\mathcal{H}) \subset \mathcal{L}_{\frac{p+1}{j+1}}(\mathcal{H}).$$

Thus,

$$(A_0 + B - z)^{-k} B(A_0 - z)^{-j-1+k} \in \mathcal{L}_{\frac{p+1}{j+1}}(\mathcal{H}), \quad k = 1, \dots, j-1.$$

Decomposing $(A_0 + B_n - z)^{-k} P_n B(A_0 - z)^{-j-1+k} P_n$ similarly and employing Lemma 3.1.3 one can show that

$$\lim_{n\to\infty} \left\| (A_0 + B_n - z)^{-k} P_n B(A_0 - z)^{-j-1+k} P_n - (A_0 + B - z)^{-k} B(A_0 - z)^{-j-1+k} \right\|_{\frac{p+1}{j+1}} = 0 \tag{3.9}$$

for $k = 0, \dots, j-1$.

Next, for $k = j$ using the resolvent identity we write,

$$\begin{aligned}&(A_0 + B - z)^{-j} B(A_0 - z)^{-1} \\ &\quad = (A_0 + B - z)^{-j+1} (A_0 - z)^{-1} B(A_0 - z)^{-1} \\ &\quad - (A_0 + B - z)^{-j} B(A_0 - z)^{-1} B(A_0 - z)^{-1} \\ &\quad = \Big((A_0 + B - z)^{-j+1} - (A_0 - z)^{-j+1}\Big)(A_0 - z)^{-1} B(A_0 - z)^{-1} \\ &\quad - (A_0 + B - z)^{-1}\Big((A_0 + B - z)^{-j+1} - (A_0 - z)^{-j+1}\Big) B(A_0 - z)^{-1} B(A_0 - z)^{-1}\end{aligned} \tag{3.10}$$

$$+ (A_0 - z)^{-j} B(A_0 - z)^{-1}$$
$$- (A_0 + B - z)^{-1}(A_0 - z)^{-j+1} B(A_0 - z)^{-1} B(A_0 - z)^{-1}.$$

The first and the second term of this decomposition can be treated as for the case $k = j - 1$ above. The third and the fourth terms belong to $\mathcal{L}_{\frac{p+1}{j+1}}(\mathcal{H})$ by (3.2) (and Hölder inequality for the latter). To prove the convergence

$$\lim_{n\to\infty} \left\| (A_0 + B_n - z)^{-j} P_n B(A_0 - z)^{-1} P_n - (A_0 + B - z)^{-j} B_+ (A_0 - z)^{-1} \right\|_{\frac{p+1}{j+1}} = 0 \quad (3.11)$$

one can decompose $(A_0 + B_n - z)^{-j} P_n B(A_0 - z)^{-1} P_n$ in the same way as (3.10) and treat each term separately.

Combining (3.7) and (3.8) together with (3.9) and (3.11) we conclude the proof.

□

Corollary 3.1.8 *Let $\mathcal{D}$ be the (massive/massless) Dirac operator on $\mathbb{R}^d$. Assuming that V is a perturbation of $\mathcal{D}$ as in Example 3.1.2, we have that*

$$(\mathcal{D} + V - z)^{-d} + (\mathcal{D} - z)^{-d} \in \mathcal{L}_1(\mathcal{H}), \quad z \in \mathbb{C} \setminus \mathbb{R}.$$

We note also the following

Lemma 3.1.9 ([CGL⁺22, Lemma 8.8]) *Assume that B is a p-relative trace-class perturbation of A_0 with $p = 2k$, $k \in \mathbb{N}$. Then with $A_1 = A_0 + B$ we have*

$$\left[(A_1^2 + 1)^{-k} - (A_0^2 + 1)^{-k}\right] \in \mathcal{L}_1(\mathcal{H}).$$

Proof We can write

$$\begin{aligned}
&(A_1^2 + 1)^{-k} - (A_0^2 + 1)^{-k} \\
&= (A_1 + i)^{-k}(A_1 - i)^{-k} - (A_0 + i)^{-k}(A_0 - i)^{-k} \\
&= \left[(A_1 + i)^{-k} - (A_0 + i)^{-k}\right]\left[(A_1 - i)^{-k} - (A_0 - i)^{-k}\right] \\
&\quad + \left[(A_1 + i)^{-k} - (A_0 + i)^{-k}\right](A_0 - i)^{-k} \qquad (3.12) \\
&\quad + (A_0 + i)^{-k}\left[(A_1 - i)^{-k} - (A_0 - i)^{-k}\right].
\end{aligned}$$

By Theorem 3.1.7 the first term on the right-hand side of (3.12) is a trace-class operator. Repeating an argument similar to the proof of Theorem 3.1.7 one can show (see [CGL⁺22, Lemma 8.6] for details) that the second and the third terms are also trace-class operators. □

Next, we proceed with the discussion of the right-hand side of the principal trace formula (1.2). Let, as before, B be a p-relative trace-class perturbation of a self-adjoint operator A_0. We introduce the straight-line path $\{A_s\}_{s\in[0,1]}$ joining A_0 and $A_0 + B$, by setting

$$A_s = A_0 + sB, \quad s \in [0, 1].$$

We also introduce the path $\{A_{s,n}\}_{s\in[0,1]}$ joining A_0 and $A_0 + B_n$ by

$$A_{s,n} = A_0 + sB_n, \quad s \in [0, 1], n \in \mathbb{N}.$$

We firstly note the following:

Remark 3.1.10 Repeating the proof of Theorem 3.1.7 replacing B and B_n by the operators sB and sB_n $s \in (0, 1]$, respectively, one can conclude that the functions

$$s \mapsto \| (A_s - z)^{-j} - (A_0 - z)^{-j} \|_{\frac{p+1}{j+1}},$$

$$s \mapsto \| \left(A_{s,n} - z\right)^{-j} - (A_0 - z)^{-j} \|_{\frac{p+1}{j+1}}$$

are continuous with respect to s and uniformly bounded with respect to $n \in \mathbb{N}$. ◇

Proposition 3.1.11 *Let B be a p-relative trace-class perturbation of a self-adjoint operator A_0 and let $A_s = A_0 + sB$, $s \in [0, 1]$. The function*

$$s \mapsto e^{-tA_s^2}B, \quad t > 0,$$

is a continuous $\mathcal{L}_1(\mathcal{H})$-valued function on $[0, 1]$. *In particular, the integral*

$$\int_0^1 \operatorname{tr}\left(e^{-tA_s^2}B\right)ds$$

is well-defined.

Proof Firstly we show that the operator $e^{-tA_s^2}B$ is a trace-class operator for any fixed $s \in [0, 1]$. Since the operator

$$(A_s + i)^{p+1}e^{-tA_s^2}$$

is bounded, it is sufficient to show that the operator $(A_s + i)^{-p-1}B$ is a trace-class operator. We may write

$$(A_s + i)^{-p-1}B = \left((A_s + i)^{-p-1} - (A_0 + i)^{-p-1}\right)B + (A_0 + i)^{-p-1}B.$$

As B is a p-relative trace-class perturbation of A_0 it follows that the second term on the right-hand side is a trace-class operator. For the first term, Theorem 3.1.7

implies the operator $(A_s + i)^{-p-1} - (A_0 + i)^{-p-1}$ is a trace-class operator. Hence, $(A_s + i)^{-p-1} B \in \mathcal{L}_1(\mathcal{H})$ for any $s \in [0, 1]$, as required.

Now, we prove that the mapping $s \mapsto e^{-tA_s^2} B$ is continuous in $\mathcal{L}_1(\mathcal{H})$-norm. Let $s_1, s_2 \in [0, 1]$. We set

$$p_0 = 2\lfloor \frac{p}{2} \rfloor + 1.$$

By Theorem 3.1.7 the operators $A = A_{s_1}$ and $B = A_{s_2}$ are p_0-resolvent comparable (in the sense of Definition 2.3.6). Therefore, by Proposition 2.3.7 we have

$$e^{-tA_{s_1}^2} B - e^{-tA_{s_2}^2} B = \sum_{j=1,2} T_{f,a_j}^{A_{s_1}, A_{s_2}} \Big((A_{s_1} - a_j i)^{-p_0} - (A_{s_2} - a_j i)^{-p_0} \Big) \cdot B, \tag{3.13}$$

where $f(x) = e^{-tx^2}$, $x \in \mathbb{R}, t > 0$.

By Remark 3.1.10 we have that

$$\Big\| (A_{s_1} - a_j i)^{-p_0} - (A_{s_2} - a_j i)^{-p_0} \Big\|_1 \to 0, \quad \text{as } s_1 - s_2 \to 0.$$

Furthermore, by Theorem 2.4.8 the double operator integral $T_{f,a_j}^{A_{s_1}, A_{s_2}}$, $j = 1, 2$, converges pointwise on $\mathcal{L}_1(\mathcal{H})$ to $T_{f,a_j}^{A_{s_1}, A_{s_1}}$, as $s_2 \to s_1$. Therefore,

$$\begin{aligned} &\Big\| T_{f,a_j}^{A_{s_1}, A_{s_2}} \Big((A_{s_1} - a_j i)^{-p_0} - (A_{s_2} - a_j i)^{-p_0} \Big) \Big\| \\ &\quad \le \Big\| \big(T_{f,a_j}^{A_{s_1}, A_{s_2}} - T_{f,a_j}^{A_{s_1}, A_{s_1}} \big) \Big((A_{s_1} - a_j i)^{-p_0} - (A_{s_2} - a_j i)^{-p_0} \Big) \Big\| \\ &\quad + \Big\| T_{f,a_j}^{A_{s_1}, A_{s_1}} \Big((A_{s_1} - a_j i)^{-p_0} - (A_{s_2} - a_j i)^{-p_0} \Big) \Big\| \\ &\quad \to 0, \quad s_1 - s_2 \to 0, \quad j = 1, 2. \end{aligned}$$

Thus, equality (3.13) implies that

$$\| e^{-tA_{s_1}^2} B - e^{-tA_{s_2}^2} B \|_1 \to 0, \quad s_1 - s_2 \to 0,$$

as required. □

To conclude this section we prove that the integral in Proposition 3.1.11 can be approximated by a similar integral with B replaced by B_n (and so A_s replaced by $A_{s,n}$).

Proposition 3.1.12 *We have that*

$$\lim_{n\to\infty}\int_0^1 \operatorname{tr}\left(e^{-tA_{s,n}^2}B_n\right)ds = \int_0^1 \operatorname{tr}\left(e^{-tA_s^2}B\right)ds. \tag{3.14}$$

Proof Let $s \in [0, 1]$ be fixed. We write

$$e^{-tA_{s,n}^2}B_n = (A_{s,n}+i)^{p+1}e^{-tA_{s,n}^2}\cdot(A_{s,n}+i)^{-p-1}B_n.$$

Since $A_{s,n} \to A_s$ in the strong resolvent sense (see Lemma 3.1.6) and the function $x \mapsto e^{-tx^2}(x+i)^{p+1}$, $x \in \mathbb{R}$, is continuous and bounded, [RS80, Theorem VIII.23] implies that $(A_{s,n}+i)^{p+1}e^{-tA_{s,n}^2} \to (A_s+i)^{p+1}e^{-tA_s^2}$ in the strong operator topology. Hence, by Lemma 3.1.3, to prove the convergence

$$\lim_{n\to\infty}\left\|e^{-tA_{s,n}^2}B_n - e^{-tA_s^2}B\right\|_1 = 0 \tag{3.15}$$

it is sufficient to show that

$$\lim_{n\to\infty}\left\|(A_{s,n}+i)^{-p-1}B_n - (A_s+i)^{-p-1}B\right\|_1 = 0. \tag{3.16}$$

To prove (3.16) we write

$$\begin{aligned}
&(A_{s,n}+i)^{-p-1}B_n \\
&= \left((A_{s,n}+i)^{-p-1} - (A_0+i)^{-p-1}\right)B_n + (A_0+i)^{-p-1}B_n \\
&= \left((A_{s,n}+i)^{-p-1} - (A_0+i)^{-p-1}\right)B_n + P_n(A_0+i)^{-p-1}BP_n
\end{aligned}$$

Theorem 3.1.7 implies that $\left((A_{s,n}+i)^{-p-1} - (A_0+i)^{-p-1}\right)$ converges to $\left((A_s+i)^{-p-1} - (A_0+i)^{-p-1}\right)$ in $\mathcal{L}_1(\mathcal{H})$. Moreover, combining the assumption that $(A_0+i)^{-p-1}B \in \mathcal{L}_1(\mathcal{H})$ with the strong operator convergence $P_n \to 1$ and with Lemma 3.1.3, we obtain

$$P_n(A_0+i)^{-p-1}BP_n \to (A_0+i)^{-p-1}B$$

in $\mathcal{L}_1(\mathcal{H})$. Hence, for every fixed $s \in [0, 1]$, the sequence $\{(A_{s,n}+i)^{-p-1}B_n\}_{n\in\mathbb{N}}$ converges to $(A_s+i)^{-p-1}B$ in $\mathcal{L}_1(\mathcal{H})$, which suffices to prove (3.16). Thus,

$$\lim_{n\to\infty}\left\|e^{-tA_{s,n}^2}B_n - e^{-tA_s^2}B\right\|_1 = 0$$

for every fixed $s \in [0, 1]$.

We claim that the sequence of functions

$$s \mapsto \|e^{-tA_{s,n}^2} B_n\|_1, \quad s \in [0,1], \quad t > 0,$$

is uniformly bounded (with respect to n) by a continuous function. Indeed, we have

$$\begin{aligned}\|e^{-tA_{s,n}^2} B_n\|_1 &\le \|(A_{s,n}+i)^{p+1} e^{-tA_{s,n}^2}\| \cdot \|(A_{s,n}+i)^{-p-1} B_n\|_1 \\ &\le \text{const}\, \|\big((A_{s,n}+i)^{-p-1} - (A_0+i)^{-p-1}\big) B_n\|_1 \\ &\quad + \text{const}\, \|(A_0+i)^{-p-1} B_n\|_1,\end{aligned}$$

where the constant is independent of s and n.

By Remark 3.1.10 the first term in the previous inequality involves a sequence of functions uniformly majorised by a continuous function. Since $B_n = P_n B P_n$, the second term is clearly uniformly majorised by the constant function

$$\text{const}\, \|(A_0+i)^{-p-1} B\|_1.$$

Thus, appealing to (3.15) and the dominated convergence theorem we infer that

$$\lim_{n\to\infty} \int_0^1 \operatorname{tr}\left(e^{-tA_{s,n}^2} B_n\right) ds = \int_0^1 \operatorname{tr}\left(e^{-tA_s^2} B\right) ds.$$

□

3.2 Main Setting and Assumptions

As we explained in the introduction we are interested in spectral flow, its relation to the Fredholm or Witten index and in its generalisation beyond the Fredholm setting, for certain paths of self-adjoint operators.

Following on from the motivating paper [GLM+11] we now introduce precisely the operators we work with. Our paths are restricted by the final Hypothesis 3.2.5. In this introductory discussion however we will work under the less restrictive Hypothesis 3.2.1.

Hypothesis 3.2.1 *Suppose $\mathcal{H}$ is a complex, separable Hilbert space.*

(*i*) *Assume A_- is self-adjoint on* $\operatorname{dom}(A_-) \subseteq \mathcal{H}$.
(*ii*) *Suppose we have a family of bounded self-adjoint operators $\{B(t)\}_{t\in\mathbb{R}} \subset \mathcal{L}(\mathcal{H})$, continuously differentiable with respect to t in the uniform operator norm, such that*

$$\|B'(\cdot)\| \in L_1(\mathbb{R}) \cap L_\infty(\mathbb{R}). \tag{3.17}$$

(*iii*) *Suppose that for some $p \in \mathbb{N} \cup \{0\}$ we have*

$$B'(t)(A_- + i)^{-p-1} \in \mathcal{L}_1(\mathcal{H}), \quad \int_{\mathbb{R}} \|B'(t)(A_- + i)^{-p-1}\|_1 dt < \infty. \tag{3.18}$$

In what follows, we always choose the smallest $p \in \mathbb{N} \cup \{0\}$ which satisfies (3.18).

Given Hypothesis 3.2.1 we introduce the family of self-adjoint operators $A(t)$, $t \in \mathbb{R}$, in $\mathcal{H}$, by

$$A(t) = A_- + B(t), \quad \operatorname{dom}(A(t)) = \operatorname{dom}(A_-), \quad t \in \mathbb{R}.$$

Writing

$$B(t) = B(t_0) + \int_{t_0}^{t} B'(s)\, ds, \quad t, t_0 \in \mathbb{R}, \tag{3.19}$$

with the convergent Bochner integral on the right-hand side, we conclude that the self-adjoint asymptotes

$$\operatorname*{n-lim}_{t \to \pm\infty} B(t) := B_\pm \in \mathcal{L}(\mathcal{H})$$

exist. In particular, purely for convenience of notation, we will make the choice

$$B_- = 0$$

in the following and also introduce the asymptote,

$$A_+ = A_- + B_+, \quad \operatorname{dom}(A_+) = \operatorname{dom}(A_-). \tag{3.20}$$

Assumption (3.17) and equality (3.19) also yield,

$$\sup_{t \in \mathbb{R}} \|B(t)\| \leq \int_{\mathbb{R}} \|B'(t)\| dt < \infty. \tag{3.21}$$

A simple application of the resolvent identity yields (with $t \in \mathbb{R}$, $z \in \mathbb{C} \backslash \mathbb{R}$)

$$(A(t) - zI)^{-1} = (A_\pm - zI)^{-1} - (A(t) - zI)^{-1}[B(t) - B_\pm](A_\pm - zI)^{-1},$$
$$\big\|(A(t) - zI)^{-1} - (A_\pm - zI)^{-1}\big\| \leq |\operatorname{Im}(z)|^{-2} \|B(t) - B_\pm\|,$$

and hence proves that

$$\underset{t\to\pm\infty}{\text{n-lim}}\,(A(t)-zI)^{-1}=(A_{\pm}-zI)^{-1},\quad z\in\mathbb{C}\backslash\mathbb{R}.$$

This is relevant to whether spectral flow between A_- and A_+ along the path $\{A(t)\}$ exists [Les05].

Repeating the argument of [GLM$^+$11, (3.49)] one can prove that

$$B_+(A_-+i)^{-1-p},\quad B(t)(A_-+i)^{-1-p}\in\mathcal{L}_1(\mathcal{H}),\tag{3.22}$$

that is, B_+ as well as the family $\{B(t)\}_{t\in\mathbb{R}}$ are p-relative trace-class perturbations with respect to A_-. In particular, results of Sect. 3.1 apply to the perturbations B_+ and $B(t)$, $t\in\mathbb{R}$, of the operator A_-.

As the next step in this section, we introduce the key technical ideas that enable us to use the old results of [GLM$^+$11] in an approximation scheme.

As in Sect. 3.1 we introduce a spectral 'cut-off' of the operator A_- by setting $P_n=\chi_{[-n,n]}(A_-)$. Recall also that $\text{s-lim}_{n\to\infty}P_n=1$.

Let $\{B(t)\}_{t\in\mathbb{R}}$ be a one parameter family of perturbations of A_- satisfying Hypothesis 3.2.1. We introduce the family $\{B_n(t)\}_{t\in\mathbb{R}}$, $n\in\mathbb{N}$, of reduced operators by setting

$$B_n(t):=P_nB(t)P_n,\quad t\in\mathbb{R},n\in\mathbb{N}.$$

In this case,

$$A_n(t):=A_-+B_n(t),\quad \text{dom}(A_n(t))=\text{dom}(A_-),\quad n\in\mathbb{N},\ t\in\mathbb{R}.$$

In particular, one concludes that

$$B_{+,n}:=\underset{t\to+\infty}{\text{n-lim}}\,B_n(t)=P_nB_+P_n,\tag{3.23}$$

and therefore for the reduced asymptotes $A_{+,n}$, constructed with the family $\{B(t)\}_{t\in\mathbb{R}}$ replaced by $\{B_n(t)\}_{t\in\mathbb{R}}$, we obtain

$$A_{+,n}:=A_-+B_{+,n}=A_-+P_nB_+P_n,\quad \text{dom}(A_{+,n})=\text{dom}(A_-).\tag{3.24}$$

We note that the equality (3.22) together with the definition of the projections P_n implies that

$$B_{+,n}=P_nB_+P_n\in\mathcal{L}_1(\mathcal{H}).$$

The following proposition shows that the family $\{B_n(t)\}_{t\in\mathbb{R}}$ of 'approximants' consists of trace-class operators, and so for this family the results of [Pus08,

GLM⁺11, CGP⁺17] hold. The proof of this proposition is a verbatim repetition of the proof of [GLM⁺11, Proposition 2.3] and is therefore omitted.

Proposition 3.2.2 *The family $\{B_n(t)\}_{t\in\mathbb{R}}$ consists of trace-class perturbations of A_- and therefore satisfies the assumption in [Pus08] and [GLM⁺11].*

We will make use of the convention mentioned in the Notations so that bold face operators $\boldsymbol{T}$ act in the Hilbert space $L_2(\mathbb{R}; \mathcal{H})$ and typically represent operators associated with a family of operators $\{T(t)\}_{t\in\mathbb{R}}$ in $\mathcal{H}$, defined by

$$\begin{aligned}
&(\boldsymbol{T}f)(t) = T(t)f(t) \ \text{ for a.e. } t \in \mathbb{R},\\
&f \in \operatorname{dom}(\boldsymbol{T}) = \big\{g \in L_2(\mathbb{R}; \mathcal{H}) \,\big|\, g(t) \in \operatorname{dom}(T(t)) \text{ for a.e. } t \in \mathbb{R}; \\
&\quad t \mapsto T(t)g(t) \text{ is (weakly) measurable; } \int_{\mathbb{R}} \|T(t)g(t)\|^2\, dt < \infty\big\}.
\end{aligned} \tag{3.25}$$

Corresponding to the problem of studying spectral flow along a suitable path joining $A_\pm$ there is an index problem that we now introduce. Let $\boldsymbol{A}_-$ be the operator acting in $L_2(\mathbb{R}; \mathcal{H})$ defined by (3.25) with a constant fibre family $\{A_-(t)\}_{t\in\mathbb{R}} = \{A_-\}_{t\in\mathbb{R}}$, that is,

$$\begin{aligned}
&(\boldsymbol{A}_-f)(t) = A_-f(t),\\
&f \in \operatorname{dom}(\boldsymbol{A}_-) = \left\{g \in L_2(\mathbb{R}; \mathcal{H}) \,\middle|\, g(t) \in \operatorname{dom}(A_-) \text{ for a.e. } t \in \mathbb{R}; \right.\\
&\quad \left. t \mapsto A_-g(t) \text{ is (weakly) measurable; } \int_{\mathbb{R}} \|A_-g(t)\|^2 dt < \infty\right\}.
\end{aligned}$$

Identifying the Hilbert spaces $L_2(\mathbb{R}; \mathcal{H})$ and $L_2(\mathbb{R}) \otimes \mathcal{H}$ we have that

$$\boldsymbol{A}_- = 1 \otimes A_-.$$

Let the operators $\boldsymbol{A}$, $\boldsymbol{B}$, $\boldsymbol{A}' = \boldsymbol{B}'$, be defined by (3.25) in terms of the families $A(t)$, $B(t)$, and $B'(t)$, $t \in \mathbb{R}$, respectively. Since $B(t)$, $B'(t)$ are bounded operators for every $t \in \mathbb{R}$ and $\|B(\cdot)\|$, $\|B'(\cdot)\| \in L_\infty(\mathbb{R})$ (see (3.21) and Hypothesis 3.2.1 (ii), respectively) we have that

$$\boldsymbol{B},\ \boldsymbol{B}' \in \mathcal{L}(L_2(\mathbb{R}; \mathcal{H})).$$

Since, in addition, $A(t) = A_- + B(t)$, we infer that

$$\boldsymbol{A} = \boldsymbol{A}_- + \boldsymbol{B}, \quad \operatorname{dom}(\boldsymbol{A}) = \operatorname{dom}(\boldsymbol{A}_-).$$

Now, to introduce the operator $\boldsymbol{D}_{\boldsymbol{A}}$ in $L_2(\mathbb{R}; \mathcal{H})$, we recall that the operator d/dt in $L_2(\mathbb{R}; \mathcal{H})$ is defined by

$$\left(\frac{d}{dt} f\right)(t) = f'(t) \text{ for a.e. } t \in \mathbb{R},$$

$$f \in \operatorname{dom}(d/dt) = W^{1,2}(\mathbb{R}; \mathcal{H}).$$

Then, the operator $\boldsymbol{D}_{\boldsymbol{A}}$ is defined by setting

$$\boldsymbol{D}_{\boldsymbol{A}} = \frac{d}{dt} + \boldsymbol{A}, \quad \operatorname{dom}(\boldsymbol{D}_{\boldsymbol{A}}) = W^{1,2}(\mathbb{R}; \mathcal{H}) \cap \operatorname{dom}(\boldsymbol{A}_-). \tag{3.26}$$

Assuming Hypothesis 3.2.1 and repeating the proof of [GLM+11, Lemma 4.4] one can show that the operator $\boldsymbol{D}_{\boldsymbol{A}}$ is densely defined and closed in $L_2(\mathbb{R}; \mathcal{H})$. Furthermore, the adjoint operator $\boldsymbol{D}_{\boldsymbol{A}}^*$ of $\boldsymbol{D}_{\boldsymbol{A}}$ in $L_2(\mathbb{R}; \mathcal{H})$ is then given by (cf. [GLM+11])

$$\boldsymbol{D}_{\boldsymbol{A}}^* = -\frac{d}{dt} + \boldsymbol{A}, \quad \operatorname{dom}(\boldsymbol{D}_{\boldsymbol{A}}^*) = W^{1,2}(\mathbb{R}; \mathcal{H}) \cap \operatorname{dom}(\boldsymbol{A}_-).$$

This enables us to introduce the non-negative, self-adjoint operators $\boldsymbol{H}_j$, $j = 1, 2$, in $L_2(\mathbb{R}; \mathcal{H})$ by

$$\boldsymbol{H}_1 = \boldsymbol{D}_{\boldsymbol{A}}^* \boldsymbol{D}_{\boldsymbol{A}}, \quad \boldsymbol{H}_2 = \boldsymbol{D}_{\boldsymbol{A}} \boldsymbol{D}_{\boldsymbol{A}}^*. \tag{3.27}$$

The following result is proved in [CGP+17, Theorem 2.6] under a relatively trace-class perturbation assumption. It was already noted in [CGP+17, Remark 2.7] that the result holds without this assumption. Thus, in our more general setting the following theorem holds.

Theorem 3.2.3 *Assume Hypothesis 3.2.1. Then the operator $\boldsymbol{D}_{\boldsymbol{A}}$ is Fredholm if and only if $0 \in \rho(A_+) \cap \rho(A_-)$.*

Next, we turn to the reduced counterparts $\boldsymbol{H}_{j,n}$, $j = 1, 2, n \in \mathbb{N}$, of the operators $\boldsymbol{H}_j$, $j = 1, 2$. Recall that the family $\{B_n(t)\}_{t\in\mathbb{R}}$, $n \in \mathbb{N}$ is defined by (see (3.23))

$$B_n(t) = P_n B(t) P_n, \quad P_n = \chi_{[-n,n]}(A_-).$$

In this case, the corresponding operator $\boldsymbol{A}_n$ is defined as

$$\boldsymbol{A}_n = \boldsymbol{A}_- + \boldsymbol{B}_n,$$

where $\boldsymbol{B}_n$ is defined by (3.25) with $\{T(t)\}_{t\in\mathbb{R}} = \{B_n(t)\}_{t\in\mathbb{R}}$.

Denote by $\boldsymbol{H}_{j,n}$, $j = 1, 2$, the operator defined by (3.27) with $\boldsymbol{D}_{\boldsymbol{A}}$ replaced by the corresponding operator $\boldsymbol{D}_{\boldsymbol{A}_n} = \frac{d}{dt} + \boldsymbol{A}_n$. Our aim is to have approximation on the left-hand side of the principal trace formula. To this end we shall use an approximation result for double operator integrals from Theorem 2.4.9. However, to use this theorem, we need to show that both $(\boldsymbol{H}_2 - z)^{-m} - (\boldsymbol{H}_1 - z)^{-m}$ and $(\boldsymbol{H}_{2,n} - z)^{-m} - (\boldsymbol{H}_{1,n} - z)^{-m}$ are trace-class and

$$\lim_{n\to\infty} \left\| \left((\boldsymbol{H}_{2,n} - z)^{-m} - (\boldsymbol{H}_{1,n} - z)^{-m} \right) - (\boldsymbol{H}_2 - z)^{-m} - (\boldsymbol{H}_1 - z)^{-m} \right\|_1 = 0.$$

To show this we introduce $\boldsymbol{H}_0$ in $L_2(\mathbb{R}; \mathcal{H})$ by

$$\boldsymbol{H}_0 = -\frac{d^2}{dt^2} + \boldsymbol{A}_-^2, \quad \operatorname{dom}(\boldsymbol{H}_0) = W^{2,2}(\mathbb{R}; \mathcal{H}) \cap \operatorname{dom}\left(\boldsymbol{A}_-^2\right).$$

By [RS80, Theorem VIII.33], the operator $\boldsymbol{H}_0$ is self-adjoint and positive. We note that the operators $\boldsymbol{A}_-$ and $\boldsymbol{H}_0$ commute and

$$\operatorname{dom} \boldsymbol{H}_0^{1/2} = \operatorname{dom}(d/dt) \cap \operatorname{dom} \boldsymbol{A}_-. \tag{3.28}$$

For $k \in \mathbb{N}$ we introduce

$$\begin{aligned}\operatorname{dom}(\delta_{\boldsymbol{H}_0}^k) = \{\boldsymbol{T} \in \mathcal{L}(L_2(\mathbb{R}; \mathcal{H})) : \boldsymbol{T} \operatorname{dom}(\boldsymbol{H}_0^{j/2}) \subset \operatorname{dom}(\boldsymbol{H}_0^{j/2}), \forall j = 1, \dots, k,\\ \text{and the operator } [(1 + \boldsymbol{H}_0)^{1/2}, \boldsymbol{T}]^{(k)}, \text{ defined on } \operatorname{dom}(\boldsymbol{H}_0^{k/2})\\ \text{extends to a bounded operator on } L_2(\mathbb{R}; \mathcal{H})\}.\end{aligned}$$

and set

$$\delta_{\boldsymbol{H}_0}^k(\boldsymbol{T}) = \overline{[(1 + \boldsymbol{H}_0)^{1/2}, \boldsymbol{T}]^{(k)}}, \quad \boldsymbol{T} \in \operatorname{dom} \delta_{\boldsymbol{H}_0}^k, \tag{3.29}$$

where the notation $[(1 + \boldsymbol{H}_0)^{1/2}, \boldsymbol{T}]^{(k)}$ stands for k-th repeated commutator defined by

$$\begin{aligned}[(1 + \boldsymbol{H}_0)^{1/2}, T]^{(k)} = [(1 + \boldsymbol{H}_0)^{1/2}, \dots [(1 + \boldsymbol{H}_0)^{1/2}, [(1 + \boldsymbol{H}_0)^{1/2}, T]] \dots],\\ \operatorname{dom}([(1 + \boldsymbol{H}_0)^{1/2}, T]^{(k)}) = \operatorname{dom}(\boldsymbol{H}_0^{k/2}).\end{aligned}$$

For convenience, we set

$$[(1 + \boldsymbol{H}_0)^{1/2}, \boldsymbol{T}]^{(0)} = \boldsymbol{T}.$$

Using methods borrowed from the abstract pseudo-differential calculus of non-commutative geometry [CGRS14, CGP+15, CM95], the following result was established in [CLPS22, Proposition 4.8 and Proposition 4.10]. This result is crucial in the proof of the convergence of the left-hand side of the principal trace formula.

Proposition 3.2.4 *Let $\boldsymbol{B}, \boldsymbol{B}' \in \bigcap_{j=1}^{k-1} \operatorname{dom}(\delta_{\boldsymbol{H}_0}^j)$ for some $k \in \mathbb{N}$ and let $z \in \mathbb{C} \setminus \mathbb{R}$. Then*

(i) The operators $\overline{(\boldsymbol{H}_i - z)^{-k/2}(\boldsymbol{H}_0 - z)^{k/2}}$, $(\boldsymbol{H}_0 - z)^{k/2}(\boldsymbol{H}_i - z)^{-k/2}$ are bounded and the sequences $\{\overline{(\boldsymbol{H}_{i,n} - z)^{-k/2}(\boldsymbol{H}_0 - z)^{k/2}}\}_{n\in\mathbb{N}}$, and $\{(\boldsymbol{H}_0 - z)^{k/2}(\boldsymbol{H}_{i,n} - z)^{-k/2}\}_{n\in\mathbb{N}}$, $i = 1, 2$, are uniformly bounded.

(ii)

$$\overline{(\boldsymbol{H}_{2,n} - z)^{-\frac{k}{2}}(\boldsymbol{H}_0 - z)^{\frac{k}{2}}} \to \overline{(\boldsymbol{H}_2 - z)^{-\frac{k}{2}}(\boldsymbol{H}_0 - z)^{\frac{k}{2}}}$$

in the strong operator topology.

(iii)

$$(\boldsymbol{H}_0 - z)^{\frac{k}{2}}(\boldsymbol{H}_{1,n} - z)^{-\frac{k}{2}} \to (\boldsymbol{H}_0 - z)^{\frac{k}{2}}(\boldsymbol{H}_1 - z)^{-\frac{k}{2}}$$

in the strong operator topology.

We now formulate the main hypothesis that we impose for the principal trace formula to hold.

Hypothesis 3.2.5

(i) Assume A_- is self-adjoint on $\operatorname{dom}(A_-) \subseteq \mathcal{H}$ with $\mathcal{H}$ a complex, separable Hilbert space.

(ii) Suppose we have a family of bounded self-adjoint operators $\{B(t)\}_{t\in\mathbb{R}} \subset \mathcal{L}(\mathcal{H})$, continuously differentiable with respect to t in the uniform operator norm, such that $\|B'(\cdot)\| \in L_1(\mathbb{R}; dt) \cap L_\infty(\mathbb{R}; dt)$.

(iii) Suppose that for some $p \in \mathbb{N} \cup \{0\}$, we have

$$B'(t)(A_- + i)^{-p-1} \in \mathcal{L}_1(\mathcal{H}), \quad \int_{\mathbb{R}} \|B'(t)(A_- + i)^{-p-1}\|_1 dt < \infty.$$

(iv) Let $m = \lceil \frac{p}{2} \rceil$. Assume that for all $z < 0$ we have that

$$\boldsymbol{B}'(\boldsymbol{H}_0 - z)^{-m-1} \in \mathcal{L}_1(L_2(\mathbb{R}; \mathcal{H})).$$

(v) $\boldsymbol{B}, \boldsymbol{B}' \in \bigcap_{j=1}^{2m-1} \operatorname{dom}(\delta_{\boldsymbol{H}_0}^j)$, where $\delta_{\boldsymbol{H}_0}$ is defined in (3.29).

In what follows we always take the smallest $p \in \mathbb{N} \cup \{0\}$ satisfying (iii).

Given Hypothesis 3.2.5 we now state the second main approximation result that is essential for the proof of the principal trace formula in Sect. 6.2 below.

Theorem 3.2.6 *Assume Hypothesis 3.2.5. Let* $z \in \mathbb{C} \setminus \mathbb{R}_+$.

(i) Both $(\boldsymbol{H}_2 - z)^{-m} - (\boldsymbol{H}_1 - z)^{-m}$ *and* $(\boldsymbol{H}_{2,n} - z)^{-m} - (\boldsymbol{H}_{1,n} - z)^{-m}$ *are trace-class.*
(ii) We have

$$\lim_{n\to\infty} \left\| \left((\boldsymbol{H}_{2,n} - z)^{-m} - (\boldsymbol{H}_{1,n} - z)^{-m}\right) - (\boldsymbol{H}_2 - z)^{-m} - (\boldsymbol{H}_1 - z)^{-m} \right\|_1 = 0.$$

Proof

(i) Hypothesis 3.2.5 (v) guarantees that the operator $\boldsymbol{A}_{-}\boldsymbol{B}$ is well defined on $\operatorname{dom} \boldsymbol{H}_0^{1/2}$, since $\boldsymbol{B} \operatorname{dom} \boldsymbol{H}_0^{1/2} \subset \operatorname{dom} \boldsymbol{H}_0^{1/2}$ and $\operatorname{dom} \boldsymbol{H}_0^{1/2} \subset \operatorname{dom} \boldsymbol{A}_{-}$ (see (3.28)). Therefore, recalling that $\boldsymbol{A} = \boldsymbol{A}_{-} + \boldsymbol{B}$ one can decompose $\boldsymbol{H}_j$, $j = 1, 2$, as follows

$$\begin{aligned} \boldsymbol{H}_j &= -\frac{d^2}{dt^2} + \boldsymbol{A}^2 + (-1)^j \boldsymbol{A}' \\ &= \boldsymbol{H}_0 + \boldsymbol{B}\boldsymbol{A}_{-} + \boldsymbol{A}_{-}\boldsymbol{B} + \boldsymbol{B}^2 + (-1)^j \boldsymbol{B}', \\ &\operatorname{dom}(\boldsymbol{H}_j) = \operatorname{dom}(\boldsymbol{H}_0), \quad j = 1, 2. \end{aligned}$$

Hence, using the resolvent identity we write

$$\begin{aligned} &(\boldsymbol{H}_2 - z)^{-m} - (\boldsymbol{H}_1 - z)^{-m} \\ &\quad = \sum_{j=1}^{m} (\boldsymbol{H}_2 - z)^{-m+j}((\boldsymbol{H}_2 - z)^{-1} - (\boldsymbol{H}_1 - z)^{-1})(\boldsymbol{H}_1 - z)^{-j+1} \\ &\quad = -2\sum_{j=1}^{m} (\boldsymbol{H}_2 - z)^{-m+j-1} \boldsymbol{B}'(\boldsymbol{H}_1 - z)^{-j}. \end{aligned}$$

Thus

$$\begin{aligned} (\boldsymbol{H}_2 - z)^{-m} - (\boldsymbol{H}_1 - z)^{-m} &= -2\sum_{j=1}^{m} \overline{(\boldsymbol{H}_2 - z)^{-m+j-1}(\boldsymbol{H}_0 - z)^{m-j+1}} \\ &\times (\boldsymbol{H}_0 - z)^{-m+j-1} \boldsymbol{B}'(\boldsymbol{H}_0 - z)^{-j} \times (\boldsymbol{H}_0 - z)^{j}(\boldsymbol{H}_1 - z)^{-j} \end{aligned} \tag{3.30}$$

Hypothesis 3.2.5 (iv) and the three lines theorem imply that

$$(\boldsymbol{H}_0 - z)^{-m+j-1} \boldsymbol{B}'(\boldsymbol{H}_0 - z)^{-j} \in \mathcal{L}_1(L_2(\mathbb{R}; \mathcal{H})) \tag{3.31}$$

for all $j = 1, \dots, m$. Since, in addition, by Proposition 3.2.4, the operators $\overline{(\boldsymbol{H}_2 - z)^{-m+j-1}(\boldsymbol{H}_0 - z)^{m-j+1}}$ and $(\boldsymbol{H}_0 - z)^{j}(\boldsymbol{H}_1 - z)^{-j}$ are bounded, we infer that $(\boldsymbol{H}_2 - z)^{-m} - (\boldsymbol{H}_1 - z)^{-m} \in \mathcal{L}_1(L_2(\mathbb{R}; \mathcal{H}))$.

Arguing similarly, one may obtain that

$$
\begin{aligned}
&(\boldsymbol{H}_{2,n} - z)^{-m} - (\boldsymbol{H}_{1,n} - z)^{-m} \\
&\quad = -2\sum_{j=1}^{m} (\boldsymbol{H}_{2,n} - z)^{-m+j-1} \boldsymbol{P}_n \boldsymbol{B}' \boldsymbol{P}_n (\boldsymbol{H}_{1,n} - z)^{-j-1} \\
&\quad = -2\sum_{j=1}^{m} \overline{(\boldsymbol{H}_{2,n} - z)^{-m+j-1} (\boldsymbol{H}_0 - z)^{m-j+1}} \\
&\qquad \times \boldsymbol{P}_n (\boldsymbol{H}_0 - z)^{-m+j-1} \boldsymbol{B}' (\boldsymbol{H}_0 - z)^{-j} \boldsymbol{P}_n \times (\boldsymbol{H}_0 - z)^{j} (\boldsymbol{H}_{1,n} - z)^{-j}.
\end{aligned}
\tag{3.32}
$$

Referring to Proposition 3.2.4 we have that $\left(\boldsymbol{H}_{2,n} - z\right)^{-m} - \left(\boldsymbol{H}_{1,n} - z\right)^{-m} \in \mathcal{L}_1(L_2(\mathbb{R}; \mathcal{H}))$.

(ii) Using decompositions (3.30) and (3.32) we see that it is sufficient to prove the convergence of each term separately.

By (3.31), the operator $(\boldsymbol{H}_0 - z)^{-m+j-1} \boldsymbol{B}' (\boldsymbol{H}_0 - z)^{-j} \in \mathcal{L}_1(L_2(\mathbb{R}; \mathcal{H}))$ for all $j = 1, \ldots, m$, and therefore, by Lemma 3.1.3 we have that

$$
\boldsymbol{P}_n (\boldsymbol{H}_0 - z)^{-m+j-1} \boldsymbol{B}' (\boldsymbol{H}_0 - z)^{-j} \boldsymbol{P}_n \xrightarrow{\|\cdot\|_1} (\boldsymbol{H}_0 - z)^{-m+j-1} \boldsymbol{B}' (\boldsymbol{H}_0 - z)^{-j}.
$$

In addition, by Proposition 3.2.4 we have

$$
\overline{(\boldsymbol{H}_{2,n} - z)^{-m+j-1} (\boldsymbol{H}_0 - z)^{m-j+1}} \to \overline{(\boldsymbol{H}_2 - z)^{-m+j-1} (\boldsymbol{H}_0 - z)^{m-j+1}}
$$

and

$$
(\boldsymbol{H}_0 - z)^{j} (\boldsymbol{H}_{1,n} - z)^{-j} \to (\boldsymbol{H}_0 - z)^{j} (\boldsymbol{H}_1 - z)^{-j}, \quad j = 1, \ldots, m,
$$

in the strong operator topology. Thus, another appeal to Lemma 3.1.3 completes the proof.

□

Next, we discuss some details of our main assumption, Hypothesis 3.2.5, for the special path $\{B(t)\}_{t\in\mathbb{R}}$ defined as follows. Suppose that a positive function θ on $\mathbb{R}$ satisfies

$$
\begin{aligned}
&\theta \in C_b^\infty(\mathbb{R}), \quad \theta' \in L_1(\mathbb{R}), \\
&\lim_{t\to-\infty} \theta(t) = 0, \quad \lim_{t\to+\infty} \theta(t) = 1.
\end{aligned}
\tag{3.33}
$$

and assume that $B_+ \in \mathcal{L}(\mathcal{H})$. Introduce then the family $\{B(t)\}_{t\in\mathbb{R}}$ given by

$$
B(t) = \theta(t) B_+. \tag{3.34}
$$

Proposition 3.2.7 *Suppose that* B_+ *is a p-relative trace-class perturbation of* A_- *and let* $\{B(t)\}_{t\in\mathbb{R}}$ *be as in* (3.34) *with* θ *satisfying* (3.33). *Then the assumptions (ii) and (iii) of Hypothesis 3.2.1 are satisfied.*

Moreover, an argument similar to the proof of [CGK16, Proposition 2.2] guarantees the following

Proposition 3.2.8 ([CGK16]) *Suppose that* B_+ *is a p-relative trace-class perturbation of* A_- *and let* $\{B(t)\}_{t\in\mathbb{R}}$ *be as in* (3.34) *with* θ *satisfying* (3.33). *Then*

$$\boldsymbol{B}'(\boldsymbol{H}_0 - z)^{-m-1} \in \mathcal{L}_1(L_2(\mathbb{R};\mathcal{H})),$$

that is, assumption (iv) of Hypothesis 3.2.5 is satisfied.

Thus, for the special type of family $\{B(t)\}_{t\in\mathbb{R}} = \{\theta(t)B_+\}_{t\in\mathbb{R}}$ assumptions (ii), (iii) and (iv) of Hypothesis 3.2.5 are automatically guaranteed by the assumption (3.33) on θ and the fact that B_+ is a p-relative trace-class perturbation of A_-.

Next, we discuss Hypothesis 3.2.5 (v). By definition of $\delta_{\boldsymbol{H}_0}$ (see (3.29)) to check Hypothesis 3.2.5 (v) we have to consider repeated commutators with $(1+\boldsymbol{H}_0)^{1/2}$. However, in general, it is hard to work with these commutators. Therefore, we use below a different type of commutator argument, in which the commutators are more manageable.

Following [CGRS14, Section 1.3] for a self-adjoint operator A on $\mathcal{H}$ we introduce the operator

$$L^k_{A^2}(T) = \overline{(1+A^2)^{-k/2}[A^2,T]^{(k)}} \tag{3.35}$$

with domain

$$\begin{aligned}\operatorname{dom}(L^k_{A^2}) = \{T \in \mathcal{L}(\mathcal{H}) : T\operatorname{dom}(A^j) \subset \operatorname{dom}(A^j),\ j = 1,\dots,2k \\ \text{and the operator } (1+A^2)^{-k/2}[A^2,T]^{(k)} \text{ defined on } \operatorname{dom}(A^{2k}) \\ \text{extends to a bounded operator on } \mathcal{H}\}.\end{aligned}$$

Proposition 3.2.9 ([CLPS22, Proposition 5.4]) *Let* $\{B(t)\}_{t\in\mathbb{R}}$ *be as in* (3.34) *with* θ *satisfying* (3.33). *If* $B_+ \in \bigcap_{j=1}^k \operatorname{dom}(L^j_{A_-^2})$, *for some* $k \in \mathbb{N}$, *then* $\boldsymbol{B}, \boldsymbol{B}' \in \bigcap_{j=1}^k \operatorname{dom}(L^j_{\boldsymbol{H}_0})$.

We now formulate the Hypothesis 3.2.5 for the special case when the family $\{B(t)\}_{t\in\mathbb{R}}$ is given by $\{\theta(t)B_+\}_{t\in\mathbb{R}}$.

Hypothesis 3.2.10

(i) Assume that A_- *is self-adjoint on* $\operatorname{dom}(A_-) \subseteq \mathcal{H}$ *with* $\mathcal{H}$ *a complex, separable Hilbert space and let* θ *satisfy* (3.33).

(ii) *Suppose that an operator B_+ is a p-relative trace-class perturbation of A_- for some $p \in \mathbb{N}$, that is*

$$B_+(A_- + i)^{-p-1} \in \mathcal{L}_1(\mathcal{H}).$$

(iii) *Assume also that $B_+ \in \bigcap_{j=1}^{2p} \operatorname{dom}(L^j_{A_-^2})$, where the mapping $L^j_{A_-^2}$ is defined by* (3.35).

The proof of the following proposition follows from a combination of Propositions 3.2.7, 3.2.8 and 3.2.9.

Proposition 3.2.11 *For the special case when $\{B(t)\}_{t\in\mathbb{R}} = \{\theta(t)B_+\}_{t\in\mathbb{R}}$, Hypothesis 3.2.10 guarantees that Hypothesis 3.2.5 is satisfied.*

Chapter 4
The Spectral Shift Function

As we noted in the introduction the underlying philosophy is that to go beyond the Fredholm case we need a new perspective on index theory. As the spectral shift function calculates spectral flow in the Fredholm case, we regard it as the natural generalisation of spectral flow to the non-Fredholm case because it remains defined without Fredholm assumptions. In this chapter we introduce the spectral shift function and some of its history.

We begin with a historical exposition of the spectral shift function going back to its origins in the work of M. G. Krein. We then discuss known results about the spectral shift function relevant to our later needs including the notion of resolvent comparable operators and generalisations following Yafaev's extensive work. Yafaev's constructions allows us to introduce the spectral shift function $\xi(\cdot; A_+, A_-)$ whenever the endpoints $A_\pm$ are not Fredholm. We pay special attention to a problem that arises in fixing an arbitrary additive constant in the definition of $\xi(\cdot; A_+, A_-)$. The constant appears naturally in Yafaev's construction. As we intend to express the Witten index of the operator $\boldsymbol{D}_A$ in terms of the spectral shift function $\xi(\cdot; A_+, A_-)$ this additive constant must be determined.

To have a discussion in Chap. 7 of our result in the setting of Dirac operators on $\mathbb{R}^d$ we describe a special representation of the spectral shift function in terms of perturbation determinants. This representation allows us to link the spectral data of the perturbed operator to the behaviour of the spectral shift function. Finally both spectral shift functions $\xi(\cdot; A_+, A_-)$ and $\xi(\cdot; \boldsymbol{H}_2, \boldsymbol{H}_1)$ are introduced and the arbitrary additive constant in $\xi(\cdot; A_+, A_-)$ is fixed.

A. Carey, G. Levitina, *Index Theory Beyond the Fredholm Case*, Lecture Notes in Mathematics 2323, https://doi.org/10.1007/978-3-031-19436-8_4

4.1 An Introduction to the Theory of the Spectral Shift Function

In this section we recall the classical construction of the spectral shift function due to M. G. Krein as well as its properties. The material of this section is based on the survey [CGLS16b]. Additional details of the theory of spectral shift function can be found in [Yaf92, Chapter 8].

In 1947, the physicist I. M. Lifshitz considered perturbations of an operator A_0 (arising as the Hamiltonian of a lattice model in quantum mechanics) by a finite-rank perturbation B and found some formulas and quantitative relations for the size of the shift of the eigenvalues. In one of his papers the spectral shift function $\xi(\,\cdot\,; A_0 + B, A_0)$ appeared for the first time, and formulas for it in the case of a finite-rank perturbation were obtained.

Lifshitz later continued these investigations and applied them to the problem of computing the trace of the operator $\varphi(A_0+B)-\varphi(A_0)$, where A_0 is the unperturbed self-adjoint operator and B is a self-adjoint, finite-dimensional perturbation, and φ is an appropriate function (belonging to a fairly broad class). He obtained (or, rather, surmised) the remarkable relation

$$\operatorname{tr}(\varphi(A_0 + B) - \varphi(A_0)) = \int_{\mathbb{R}} \varphi'(\lambda)\xi(\lambda; A_0 + B, A_0)\, d\lambda, \tag{4.1}$$

where the function $\xi(\,\cdot\,; A_0 + B, A_0)$ depends on operators A_0 and B only.

A physical example treated by Lifshitz is the following: if A_0 is the operator describing the oscillations of a crystal lattice, then the free energy of the oscillations can be represented in the form $F = \operatorname{tr}(\varphi(A_0))$, for some φ. In this case, the trace formula enables one to compute the change in the free energy of oscillations of the crystal lattice upon introduction of a foreign admixture into the crystal.

If one wants to study continuous analogues of lattice models, then the perturbations B, are no longer typically described by finite-rank operators. For such models the appropriate class of perturbations, such that the spectral shift function may be defined, needs to be specified. In his paper [Kre53], M. G. Krein introduced a suitable class of perturbations for which the spectral shift function exists. He also described the broad class of functions φ for which (4.1) holds. The formula (4.1) is customarily referred to as Lipschitz-Krein trace formula. Krein's approach of construction of the spectral shift function was based on the notion of perturbation determinants to be discussed next.

4.1.1 Perturbation Determinants

The trace-class ideal $\mathcal{L}_1(\mathcal{H})$, besides carrying the standard trace, also gives rise to the notion of a determinant, which generalizes the corresponding notion in the finite-dimensional case. Let $T \in \mathcal{L}_1(\mathcal{H})$. For any orthonormal basis $\{\omega_n\}_{n\in\mathbb{N}}$ in $\mathcal{H}$ consider the $N \times N$ matrix $\mathcal{T}_N$ with elements $\delta_{m,n} + (T\omega_m, \omega_n)$, $m, n \in 1, \ldots, N$. Then the following limit exists

$$\lim_{N\to\infty} \det(1 + \mathcal{T}_N) =: \det(1 + T), \tag{4.2}$$

independently of the choice of the basis $\{\omega_n\}_{n\in\mathbb{N}}$ (cf., [GK69, Ch. IV]). The functional $\det(1 + \cdot) : \mathcal{L}_1(\mathcal{H}) \to \mathbb{C}$ is called the *determinant*; it is continuous with respect to the $\mathcal{L}_1(\mathcal{H})$-norm.

In terms of eigenvalues of $T \in \mathcal{L}_1(\mathcal{H})$, $\{\lambda_k(T)\}_{k\in\mathcal{I}}$, $\mathcal{I} \subseteq \mathbb{N}$, an appropriate index set, one has

$$\det(1 + T) = \prod_{k\in\mathcal{I}} (1 + \lambda_k(T)),$$

where the product converges absolutely (due to the fact that $\sum_{k\in\mathcal{I}} |\lambda_k| < \infty$). We note the following properties of the determinant [GK69]

$$\det(1 + T^*) = \overline{\det(1 + T)}, \quad T \in \mathcal{L}_1(\mathcal{H})$$

$$\det(1 + T_1)(1 + T_2) = \det(1 + T_1)\det(1 + T_2), \quad T_1, T_2 \in \mathcal{L}_1(\mathcal{H})$$

$$\det(1 + T_1T_2) = \det(1 + T_2T_1), \quad T_1, T_2 \in \mathcal{L}(\mathcal{H}),\ T_1T_2, T_2T_1 \in \mathcal{L}_1(\mathcal{H}).$$

In the following, let A_0, A be self-adjoint operators in $\mathcal{H}$ with $\mathrm{dom}(A_0) = \mathrm{dom}(A)$, and let $B = A - A_0$. Assume that $BR_z(A_0) \in \mathcal{L}_1(\mathcal{H})$, where $R_z(T) = (T - z)^{-1}$ is the resolvent of an operator T, $z \in \rho(T)$. Under these assumptions we define the *perturbation determinant*

$$\Delta(z) = \Delta_{A/A_0}(z) := \det(1 + BR_z(A_0)) = \det\left((A - z)(A_0 - z)^{-1}\right), \quad \mathrm{Im}(z) \neq 0.$$

We briefly recall some properties of perturbation determinants (see [Yaf92, Chapter 8, Section 1]). For self-adjoint operators A_0, A the mapping $z \to \Delta_{A/A_0}(z)$ is analytic in both the half-planes $\mathrm{Im}(z) > 0$ and $\mathrm{Im}(z) < 0$ and

$$\Delta_{A/A_0}(\bar{z}) = \overline{\Delta_{A/A_0}(z)}, \quad \mathrm{Im}(z) \neq 0.$$

One has

$$\Delta_{A/A_0}(z) \neq 0, \quad \mathrm{Im}(z) \neq 0. \tag{4.3}$$

In addition, since $B \in \mathcal{L}_1(\mathcal{H})$, standard properties of resolvents imply that

$$\|BR_{A_0}(z)\|_1 \to 0 \text{ as } |\mathrm{Im}(z)| \to \infty,$$

and therefore,

$$\Delta_{A/A_0}(z) \to 1 \text{ as } |\mathrm{Im}(z)| \to \infty.$$

As the function $\Delta_{A/A_0}(\cdot)$ is analytic in the open upper and lower half plane and because $\Delta_{A/A_0}(z) \neq 0$, $\mathrm{Im}(z) \neq 0$, it is a standard fact from complex analysis that there exists a function $G(\cdot)$ analytic in both of the upper and lower half planes such that $e^G = \Delta_{A/A_0}$. It is natural to use the notation $\ln(\Delta_{A/A_0})$ for G. Clearly, the function $\ln(\Delta_{A/A_0})$ is multivalued and its different values at a point z, $\mathrm{Im}(z) \neq 0$, differ by $2\pi ik$, $k \in \mathbb{Z}$. Since $\Delta_{A/A_0}(z) \to 1$, as $|\mathrm{Im}(z)| \to \infty$, one *fixes the branch* of the function $\ln(\Delta_{A/A_0})$ by requiring that $\ln(\Delta_{A/A_0}(z)) \to 0$ as $|\mathrm{Im}(z)| \to \infty$.

4.1.2 M. G. Krein' s Construction of the Spectral Shift Function

To construct the spectral shift function by Krein's method we exploit the following representation of the function $\ln(\Delta_{A/A_0}(z))$,

$$\ln(\Delta_{A/A_0}(z)) = \int_{\mathbb{R}} \frac{\xi(\lambda; A, A_0)\, d\lambda}{\lambda - z}, \quad \mathrm{Im}(z) \neq 0, \tag{4.4}$$

with a real-valued $\xi(\,\cdot\,; A, A_0) \in L_1(\mathbb{R})$.

The proof of (4.4) relies on the following classical result from complex analysis.

Theorem 4.1.1 (Privalov Representation Theorem) *Suppose that F is holomorphic in the open upper half-plane. If* $\mathrm{Im}(F)$ *is bounded and non-negative (respectively, non-positive) and if* $\sup_{y\geq 1} y|F(iy)| < \infty$, *then there exists a nonnegative (respectively, non-positive) real-valued function* $\xi \in L_1(\mathbb{R})$ *such that*

$$F(z) = \int_{\mathbb{R}} \frac{\xi(\lambda)\, d\lambda}{z - \lambda}, \quad \mathrm{Im}(z) > 0.$$

The function ξ is uniquely determined by the Stieltjes inversion formula,

$$\xi(\lambda) = \frac{1}{\pi} \lim_{\varepsilon \downarrow 0+} \mathrm{Im}(F(\lambda + i\varepsilon)) \text{ for a.e. } \lambda \in \mathbb{R}.$$

We now sketch the proof of the first theorem of Krein (see Theorem 4.1.2). To verify the assumptions in Privalov's Theorem for $F = \ln(\Delta_{A/A_0})$, Krein proceeded as follows.

First, suppose that $\mathrm{rank}(B) = 1$, that is, $B = \gamma(\cdot, h)h$, $h \in \mathcal{H}$, $\|h\| = 1$, $\gamma \in \mathbb{R}$. Then

$$\Delta_{A/A_0}(z) = 1 + \gamma(R_{A_0}(z)h, h).$$

Using this explicit form of the perturbation determinant one can prove that the function $\ln(\Delta_{A/A_0}(\cdot))$ satisfies all the assumptions in Privalov's theorem (for details see, e.g., [Yaf92]). Hence, there exists a function $\xi(\lambda; A, A_0)$ satisfying (4.4), and furthermore, the function $\xi(\cdot; A, A_0)$ can be expressed in the form

$$\xi(\lambda; A, A_0) = \frac{1}{\pi} \lim_{\varepsilon \to +0} \operatorname{Im}(\ln(\Delta_{A/A_0}(\lambda + i\varepsilon))), \quad \text{a.e. } \lambda \in \mathbb{R}. \tag{4.5}$$

Suppose now, that $\operatorname{rank}(B) = n < \infty$, that is,

$$B = \sum_{k=1}^{n} \gamma_k(\cdot, h_k)h_k, \quad \gamma_k = \bar{\gamma}_k, \ \|h_k\|_1, \ 1 \le k \le n.$$

Using the notation

$$B_m = \sum_{k=1}^{m} \gamma_k(\cdot, h_k)h_k, \quad A_m = A_0 + B_m, \quad 1 \le m \le \operatorname{rank}(B),$$

one infers that the difference $A_m - A_{m-1}$ is a rank-one operator. In addition, by the multiplicative property of the determinant one concludes that

$$\ln(\Delta_{A/A_0}(z)) = \sum_{m=1}^{n} \ln(\Delta_{A_m/A_{m-1}}(z)). \tag{4.6}$$

Applying the first step to the operators A_m, A_{m-1} one infers the existence of the corresponding spectral shift functions $\xi(\cdot; A_m, A_{m-1})$, $1 \le m \le \operatorname{rank}(B)$.

Set

$$\xi(\lambda; A, A_0) = \sum_{m=1}^{n} \xi(\lambda; A_m, A_{m-1}), \quad 1 \le k \le n.$$

There are $L_1(\mathbb{R})$-norm estimates for each $\xi(\cdot; A_m, A_{m-1})$ which ensure that the function $\xi(\lambda; A, A_0)$ is integrable. Furthermore, since for every m, the representations (4.4) and (4.5) for $\ln(\Delta_{A_m/A_{m-1}})$ and $\xi(\lambda; A_m, A_{m-1})$, respectively, hold, one can infer from (4.6) and the definition of $\xi(\lambda; A, A_0)$ that representations (4.4) and (4.5) hold also for $\ln(\Delta_{A/A_0})$ and $\xi(\lambda; A, A_0)$.

Suppose now, that B is an arbitrary trace-class perturbation. Let B_n be a sequence of finite-rank operators, such that $\|B - B_n\|_1 \to 0$, $n \to \infty$. Set

$$\xi(\lambda; A, A_0) = \sum_{n} \xi(\lambda; A_n, A_{n-1}),$$

where the sum now is infinite (unless, B is a finite-rank operator).

Then the convergence properties of determinants and the $L_1(\mathbb{R})$-norm estimate for each $\xi(\,\cdot\,; A_n, A_{n-1})$ imply that this series converges in $L_1(\mathbb{R})$ and all the desired representations (4.4) and (4.5) for $\ln(\Delta_{A/A_0})$ and $\xi(\,\cdot\,; A, A_0)$ hold.

The following result is the first theorem of M. G. Krein.

Theorem 4.1.2 ([Kre53]) *Let $B \in \mathcal{L}_1(\mathcal{H})$ be self-adjoint. Then the following representation holds*

$$\ln(\Delta_{A/A_0}(z)) = \int_{\mathbb{R}} \frac{\xi(\lambda; A, A_0)\, d\lambda}{\lambda - z}, \quad \operatorname{Im}(z) \neq 0,$$

where

$$\xi(\lambda; A, A_0) = \frac{1}{\pi} \lim_{\varepsilon \to 0+} \operatorname{Im}(\ln(\Delta_{A/A_0}(\lambda + i\varepsilon))) \text{ for a.e. } \lambda \in \mathbb{R}, \tag{4.7}$$

in particular, the limit in (4.7) *exists for a.e.* $\lambda \in \mathbb{R}$. *In addition,*

$$\int_{\mathbb{R}} |\xi(\lambda; A, A_0)|\, d\lambda \le \|B\|_1, \qquad \int_{\mathbb{R}} \xi(\lambda; A, A_0)\, d\lambda = \operatorname{tr}(B).$$

Moreover, $\xi(\lambda; A, A_0) \le k_+$ (respectively, $\xi(\lambda; A, A_0) \ge -k_-$) for a.e. $\lambda \in \mathbb{R}$, provided that the perturbation B has only k_+ positive (respectively, k_- negative) eigenvalues.

For the spectral shift function constructed in Theorem 4.1.2, Krein proved the Lifshitz–Krein trace formula.

Theorem 4.1.3 (Second Theorem of M. G. Krein) *Let $B \in \mathcal{L}_1(\mathcal{H})$ and assume that $f \in C^1(\mathbb{R})$ and its derivative admits the representation*

$$f'(\lambda) = \int_{\mathbb{R}} \exp(-i\lambda t)\, dm(t), \quad |m|(\mathbb{R}) < \infty,$$

for a finite (complex) measure m. Then $[f(A) - f(A_0)] \in \mathcal{L}_1(\mathcal{H})$, and the following trace formula holds

$$\operatorname{tr}(f(A) - f(A_0)) = \int_{\mathbb{R}} f'(\lambda)\xi(\lambda; A, A_0)\, d\lambda. \tag{4.8}$$

Remark 4.1.4

(i) The function $\xi(\,\cdot\,; A, A_0)$ is an element of $L_1(\mathbb{R})$, that is, it represents an *equivalence class* of Lebesgue measurable functions. Therefore, generally speaking, the notation $\xi(\lambda; A, A_0)$ is meaningless for a fixed $\lambda \in \mathbb{R}$.

(ii) For a trace-class perturbation B, the spectral shift function $\xi(\,\cdot\,; A, A_0)$ is *unique*. For more general perturbation this is no longer the case, in general. See the discussion in Sect. 4.2.

(iii) The Lifshitz–Krein trace formula (4.8) can be extended in various ways. One could attempt to describe the class of functions f, for which this formula holds; however, we will not cover this direction. Another important direction is to enlarge the class of perturbations $A - A_0$. This direction is of primary interest to us and we will present some results in Sect. 4.2. ◇

4.1.3 Properties of the Spectral Shift Function

Let A_0, A_1 and A be such that $(A_1 - A_0), (A - A_1) \in \mathcal{L}_1(\mathcal{H})$. Then we have for a.e. $\lambda \in \mathbb{R}$

$$\xi(\lambda; A, A_1) + \xi(\lambda; A_1, A_0) = \xi(\lambda; A, A_0),$$

in particular, $\xi(\lambda; A, A_0) = -\xi(\lambda; A_0, A)$, and also the inequality

$$\|\xi(\cdot\,; A, A_0) - \xi(\cdot\,; A_1, A_0)\|_1 \leq \|A - A_1\|_1$$

holds. In addition, if $A \geq A_1$, then

$$\xi(\lambda; A, A_0) \geq \xi(\lambda; A_1, A_0) \text{ for a.e. } \lambda \in \mathbb{R}.$$

Next, we describe some special situations where one can select concrete representatives from the equivalence class $\xi(\cdot\,; A, A_0)$, which justifies the term "the spectral shift function". These properties of the spectral shift function $\xi(\cdot\,; A, A_0)$ are associated with the spectra of the operators A_0 and A. For the complete proof we refer to [Yaf92, Ch. 8]

(i) Let δ be an interval (possibly unbounded) such that $\delta \subset \rho(A_0) \cap \rho(A)$. Then $\xi(\cdot\,; A, A_0)$ takes a *constant integer value* on δ, that is,

$$\xi(\lambda; A, A_0) = n, \quad n \in \mathbb{Z},\ \lambda \in \delta.$$

If the interval δ contains a half-line, then the integrability condition on ξ implies that $n = 0$.

(ii) Let μ be an isolated eigenvalue of multiplicity $\alpha_0 < \infty$ of A_0 and multiplicity α for A. Then

$$\xi(\mu_+; A, A_0) - \xi(\mu_-; A, A_0) = \alpha_0 - \alpha. \tag{4.9}$$

Property (ii) can be generalized as follows:

(iii) Suppose that in some interval (a_0, b_0) the spectrum of A_0 is discrete (i.e., the spectrum of A_0 consists at most of eigenvalues of A_0 of *finite* multiplicity all of which are *isolated* points of $\sigma(A_0)$). Then, by Weyl's theorem on the invariance of essential spectra, A has discrete spectrum in (a_0, b_0) as well. Let $\delta = (a, b)$, $a_0 < a < b < b_0$. Introduce the *eigenvalue counting functions* $N_{A_0}(\delta)$ and $N_A(\delta)$ of the operators A_0 and A, respectively, in the interval δ as the sum of the multiplicities of the eigenvalues in δ of the operator A_0, respectively, A. Since the interval δ is finite and both operators A_0, A have discrete spectrum, $N_{A_0}(\delta)$ and $N_A(\delta)$ are finite. In this case one has the equality,

$$\xi(b_-; A, A_0) - \xi(a_+; A, A_0) = N_{A_0}(\delta) - N_A(\delta). \tag{4.10}$$

The preceding property implies, in particular, the following fact.

(iv) Let A_0 be a nonnegative self-adjoint operator with purely discrete spectrum (i.e., $\sigma_{ess}(A_0) = \emptyset$). Since the perturbation B is trace-class, there exists $c \in \mathbb{R}$, such that $A \geq c$, that is, A is also lower semibounded. Generally, A will of course not be nonnegative and so one should expect negative eigenvalues of A. Thus, property (iii) implies that for $\lambda < 0$,

$$\xi(\lambda_-; A_0, A) = -N_A(-\infty, 0).$$

On the other hand, the following result demonstrates that any function from $L_1(\mathbb{R})$ arises as the spectral shift function for some pair of operators.

(v) Let ξ be an arbitrary real-valued element of $L_1(\mathbb{R})$. Then, there exists a pair of self-adjoint operators A_0, A, such that $(A - A_0) \in \mathcal{L}_1(\mathcal{H})$ and ξ is the spectral shift function $\xi(\cdot\,; A, A_0)$ for the pair (A, A_0). In addition, if $0 \leq \xi \leq 1$, then there is a pair A_0, A such that $A - A_0$ is a positive rank-one operator [Kre53, Kre81].

4.2 More General Classes of Perturbations

In this section we consider the situation where the perturbation is no longer a trace-class operator.

The first result, generalising the class of operators A_0, A for which the spectral shift function is well-defined is due to M. G. Krein [Kre62]. This result guarantees existence of the spectral shift function of resolvent comparable operators, that is for self-adjoint operators A, A_0 such that

$$(A - z)^{-1} - (A_0 - z)^{-1} \in \mathcal{L}_1(\mathcal{H}), \quad z \in \rho(A_0) \cap \rho(A).$$

In this case, the spectral shift function for the pair (A, A_0) is defined via the Cayley transform and the spectral shift function for unitary operators. For this reason we firstly present the construction of the spectral shift function for a pair of unitary operators.

4.2.1 Spectral Shift Function for Unitary Operators

We consider unitary operators U, U_0, such that $U - U_0 \in \mathcal{L}_1(\mathcal{H})$. We refer the reader to [Yaf92, Section 8.5] for the full exposition of this topic.

We denote by $\mathbb{D}$ the open unit disk in $\mathbb{C}$ and by $\mathbb{T}$ the unit circle in $\mathbb{C}$, using the parametrization

$$\zeta = e^{it}, \quad t \in [0, 2\pi],$$

for $\zeta \in \mathbb{T}$. In addition, we denote the normalized Lebesgue measure on $\mathbb{T}$ by

$$d\mu_0(\zeta) = \frac{dt}{2\pi}, \quad \zeta = e^{it},\ t \in [0, 2\pi].$$

The trace formula for the pair of unitary operators (U, U_0) then reads

$$\operatorname{tr}\big(g(U) - g(U_0)\big) = \oint_{\mathbb{T}} g'(\zeta)\eta(\zeta; U, U_0)\, d\zeta, \quad g \in C^\infty(\mathbb{T}).$$

The real-valued function $\eta(\cdot; U, U_0)$ is called the spectral-shift function for the pair (U, U_0). The important difference between the unitary case and the case of a trace-class perturbation for self-adjoint operators, is that the spectral shift function $\eta(\cdot; U, U_0)$ is no longer unique. Indeed, $\eta(\cdot; U, U_0)$ is defined only up to an additive (integer-valued) constant.

Let U and U_0 be unitary operators on some Hilbert space $\mathcal{H}$ and let

$$B := [U - U_0] \in \mathcal{L}_1(\mathcal{H}).$$

Denote by $\Delta_{U/U_0}(\cdot)$ the perturbation determinant for the pair (U, U_0), that is,

$$\Delta_{U/U_0}(z) := \det\big((U - z)(U_0 - z)^{-1}\big) = \det\big(1 + B(U_0 - z)^{-1}\big), \quad z \in \mathbb{C}\backslash\mathbb{T}.$$

As in the case of self-adjoint operators (cf., e.g., [Yaf92, Section 8.1, p. 266]) we have that

$$\Delta_{U/U_0}(z) \neq 0, \quad z \in \mathbb{C}\backslash\mathbb{T}, \tag{4.11}$$

the function $\Delta_{U/U_0}(\cdot)$ is analytic on $\mathbb{C}\setminus\mathbb{T}$, and

$$\frac{\Delta'_{U/U_0}(z)}{\Delta_{U/U_0}(z)} = \operatorname{tr}\left((U_0 - z)^{-1} - (U - z)^{-1}\right), \quad z \in \mathbb{C}\setminus\mathbb{T}. \tag{4.12}$$

In addition, (see [Yaf92, Section 8.1, Lemma 3])

$$\lim_{|z|\to\infty} \Delta_{U/U_0}(z) = 1. \tag{4.13}$$

The existence of the spectral shift function $\eta(\,\cdot\,; U, U_0)$ is proved in [Kre62] using the logarithm of the perturbation determinant, $\ln(\Delta_{U/U_0}(\cdot))$. We recall (cf., e.g., [Pal91, Theorem B.4.1]) that if $f(\cdot)$ is an analytic function free of zeros in some domain $D \subset \mathbb{C}$, then a branch of $\ln(f(\cdot))$ exists in D if and only if

$$\int_\gamma \frac{f'(z)}{f(z)} dz = 0$$

for every closed piecewise smooth path γ in D. By (4.11) the function $\Delta_{U/U_0}(\cdot)$ is analytic and free of zeros in both the exterior and interior of $\mathbb{T}$, and by (4.12) the logarithmic derivative $\Delta'_{U/U_0}(z)/\Delta_{U/U_0}(z)$ is given by $\operatorname{tr}\left((U_0 - z)^{-1} - (U - z)^{-1}\right)$. Since the function

$$z \mapsto \operatorname{tr}\left((U_0 - z)^{-1} - (U - z)^{-1}\right) \tag{4.14}$$

is analytic in the interior of the circle, $\ln(\Delta_{U/U_0}(\cdot))$ is well-defined in the interior of the circle. For the exterior of the circle the function in (4.14) has only one singularity at $z = \infty$. One can show the point $z = \infty$ is a removable singularity for this function and

$$\begin{aligned}
&\left|\operatorname{res}_{z=\infty} \operatorname{tr}\left((U_0 - z)^{-1} - (U - z)^{-1}\right)\right| \\
&\quad = \left|\lim_{z\to\infty} z \operatorname{tr}\left((U_0 - z)^{-1} - (U - z)^{-1}\right)\right| \\
&\quad \le \lim_{z\to\infty} \|z(U_0 - z)^{-1}\|_{\mathcal{L}(\mathcal{H})} \|U - U_0\|_1 \|(U - z)^{-1}\|_{\mathcal{L}(\mathcal{H})} = 0.
\end{aligned} \tag{4.15}$$

Therefore,

$$\int_\gamma \Delta'_{U/U_0}(z)/\Delta_{U/U_0}(z) dz = \int_\gamma \operatorname{tr}\left((U_0 - z)^{-1} - (U - z)^{-1}\right) dz = 0 \tag{4.16}$$

for any closed piecewise smooth path γ in the exterior of the circle. Thus, $\ln(\Delta_{U/U_0}(\cdot))$ is well-defined in the exterior of the circle also.

The condition (4.13) allows one to fix the branch of $\ln(\Delta_{U/U_0}(\cdot))$ in the exterior of $\mathbb{T}$ by requiring that

$$\lim_{|z|\to\infty} \ln(\Delta_{U/U_0}(z)) = 0.$$

In the interior of $\mathbb{T}$, one cannot naturally fix the branch of $\ln(\Delta_{U/U_0}(\cdot))$, and all branches of $\ln(\Delta_{U/U_0}(\cdot))$ differ by $2\pi i n$, $n \in \mathbb{Z}$. This is the primary reason why the spectral shift function for a pair of unitary operators is defined only up to an additive integer.

The following theorem is an analogue of Theorem 4.1.2 for unitary operators.

Theorem 4.2.1 *Assume that U, U_0 are unitary operators, such that $U - U_0 \in \mathcal{L}_1(\mathcal{H})$. The representation*

$$\ln(\Delta_{U/U_0}(z)) = \oint_{\mathbb{T}} \frac{\eta(\zeta; U, U_0)\, d\zeta}{\zeta - z}, \quad z \in \mathbb{C}\setminus\mathbb{T} \tag{4.17}$$

holds, where $\eta(\cdot; U, U_0) \in L_1(\mathbb{T})$ is defined via

$$\eta(\mu; U, U_0) = \pi^{-1} \lim_{r\to 1+0} \operatorname{Im}(\ln(\Delta_{U/U_0}(r\mu))) - (2\pi i)^{-1} \ln(\Delta_{U/U_0}(0)) \quad a.e.\, \mu \in \mathbb{T}.$$

Note that the change of the branch of $\ln(\Delta_{U/U_0}(\cdot))$ by $2\pi i n$, $n \in \mathbb{Z}$, corresponds to the change

$$\eta(\,\cdot\,; U, U_0) \to \eta(\,\cdot\,; U, U_0) + n.$$

In contrast to the self-adjoint case with a trace-class perturbation, the integrability condition $\eta(\cdot; U, U_0) \in L_1(\mathbb{T})$ does not allow us to fix the additive constant.

We now sketch the proof of Theorem 4.2.1. As before, the proof is based on the following result from complex analysis (cf., e.g., [Yaf92, Section 1.2, Corollary 13]).

Theorem 4.2.2 *Suppose that a function $F(\cdot)$ is analytic in $\mathbb{D}$, has bounded imaginary part, and* $\operatorname{Re}(F(0)) = 0$. *Then there exists a real function $\eta \in L_1(\mathbb{T}, d\mu_0)$ such that*

$$F(z) = \oint_{\mathbb{T}} \frac{\eta(\zeta)\, d\zeta}{\zeta - z}, \quad z \in \mathbb{D}.$$

In addition,

$$\eta(\zeta) = \pi^{-1}\operatorname{Im}F(\zeta_+) - (2\pi i)^{-1} F(0) \text{ for a.e. } \zeta \in \mathbb{T}, \tag{4.18}$$

where $F(\zeta_+)$ denotes the angular limit value as z tends to $\zeta \in \mathbb{T}$ from within the open unit disk $\mathbb{D}$.

Fix a branch of $\ln(\Delta_{U/U_0}(\cdot))$ in $z \in \mathbb{C}\backslash\overline{\mathbb{D}}$ by requiring that $\ln(\Delta_{U/U_0}(z)) \to 0$ as $|z| \to \infty$. For $z \in \mathbb{D}$ we fix an arbitrary branch of $\ln(\Delta_{U/U_0}(\cdot))$. For any such branch, one infers

$$\ln\big(\Delta_{U/U_0}(\bar{z}^{-1})\big) = \overline{\ln(\Delta_{U/U_0}(z))} + \ln(\Delta_{U/U_0}(0)), \quad z \in \mathbb{D}.$$

As in Sect. 4.1 the representation (4.17) is firstly proved for a rank-one perturbation $B = U - U_0$ and then extended to an arbitrary trace-class perturbation by approximation. Here we discuss the construction for a rank-one perturbation only.

Assume that the perturbation $B = U - U_0$ is a rank-one operator. Then, the operator $N := BU_0^{-1}$ may be expressed as

$$N = \gamma\langle\,\cdot\,, \psi\rangle\psi, \quad \|\psi\| = 1,$$

for some

$$\gamma = e^{i\theta} - 1, \quad \theta \in (-\pi, \pi].$$

(One notes that θ is the eigenvalue of the operator $K = \ln(N)$.)

By the definition of the perturbation determinant and the determinant,

$$\begin{aligned}
\Delta_{U/U_0}(z) &= \det\big(1 + B(U_0 - z)^{-1}\big) \\
&= \det\big(1 + BU_0^{-1}U_0(U_0 - z)^{-1}\big) \\
&= \det\big(1 + (\gamma\langle\,\cdot\,, \psi\rangle\psi)U_0(U_0 - z)^{-1}\big) \\
&= \det\big(1 + \gamma\langle\,\cdot\,, (U_0 - \bar{z})^{-1}U_0^*\psi\rangle\psi\big) \\
&= 1 + \gamma\langle U_0(U_0 - z)^{-1}\psi, \psi\rangle.
\end{aligned} \tag{4.19}$$

Introducing the auxiliary function

$$\Psi(z) = \langle (U_0 + z)(U_0 - z)^{-1}\psi, \psi\rangle, \quad z \in \mathbb{D},$$

it is clear that

$$\operatorname{Re}(\Psi(z)) = (1 - |z|^2)\|(U_0 - z)^{-1}\psi\|_{\mathcal{H}}^2 > 0. \tag{4.20}$$

Using this function, one can rewrite (4.19) as

$$\begin{aligned}
\Delta_{U/U_0}(z) &= 1 + (\gamma/2)\langle (U_0 - z + U_0 + z)(U_0 - z)^{-1}\psi, \psi\rangle \\
&= 1 + (\gamma/2) + (\gamma/2)\Psi(z), \quad z \in \mathbb{D}.
\end{aligned}$$

Using now the equality $\gamma = e^{i\theta} - 1$ and trigonometric identities we can write

$$\Delta_{U/U_0}(z) = e^{i\theta/2}\big(\cos(\theta/2) + i\sin(\theta/2)\Psi(z)\big), \quad z \in \mathbb{D}.$$

From (4.19), one infers $\Delta_{U/U_0}(0) = 1 + \gamma = e^{-i\theta}$, and therefore it is possible to choose a branch of $\ln(\Delta_{U/U_0}(\cdot))$ by requiring that

$$\operatorname{Im}(\ln(\Delta_{U/U_0}(0))) = \arg(\Delta_{U/U_0}(0)) = \theta \in (-\pi, \pi].$$

In particular,

$$\operatorname{Re}(\ln(\Delta_{U/U_0}(0))) = 0.$$

We introduce $\varphi(z) = e^{-i\theta/2}\Psi(z)$. Then

$$\arg(\varphi(z)) = \arg(\Delta_{U/U_0}(z)) - (\theta/2) = \arg(\Delta_{U/U_0}(z)) - (1/2)\arg(\Delta_{U/U_0}(0)),$$
$$\arg(\varphi(0)) = \theta/2, \tag{4.21}$$

and

$$\operatorname{Im}(\varphi(z)) = \sin(\theta/2)\operatorname{Re}(\Psi(z)).$$

Since $\operatorname{Re}(\Psi(z)) > 0$ (see (4.20)), one infers

$$\operatorname{Im}(\varphi(z)) > 0 \text{ if } \theta \in (0, \pi] \quad (\text{resp., } \operatorname{Im}(\varphi(z)) < 0 \text{ if } \theta \in (-\pi, 0)),$$

which implies that

$$0 < \arg(\varphi(z)) < \pi \text{ if } \theta \in (0, \pi] \quad (\text{resp., } -\pi < \arg(\varphi(z)) < 0 \text{ if } \theta \in (-\pi, 0)).$$

Appealing now to (4.21), one concludes that $\arg(\Delta_{U/U_0}(\cdot)) = \operatorname{Im}(\ln(\Delta_{U/U_0}(\cdot)))$ is bounded.

Thus, the function $\ln(\Delta_{U/U_0}(\cdot))$ satisfies the assumptions of Theorem 4.2.2, and there exists $\eta(\,\cdot\,; U, U_0) = \overline{\eta(\,\cdot\,; U, U_0)} \in L_1(\mathbb{T}, d\mu_0)$, such that the representation (4.17) holds. Further, by the inversion formula (4.18), one obtains

$$\eta(\mu, U, U_0) = \pi^{-1}\operatorname{Im}\ln(\Delta_{U/U_0}(\mu_+)) - (2\pi i)^{-1}\ln(\Delta_{U/U_0}(0)),$$

as required.

For the spectral shift function $\eta(\cdot; U, U_0)$ the Lipschitz-Krein trace formula holds.

Theorem 4.2.3 ([Yaf92, Theorem 8.5.6]) *Let U, U_0 be unitary operators, such that $U - U_0 \in \mathcal{L}_1(\mathcal{H})$ and assume that g is a continuously differentiable such that the Fourier series of $g'(\mu)$, $\mu \in \mathbb{T}$, converges absolutely. Then $g(U) - g(U_0) \in \mathcal{L}_1(\mathcal{H})$ and*

$$\operatorname{tr}\big(g(U) - g(U_0)\big) = \oint_{\mathbb{T}} g'(\zeta)\eta(\zeta; U, U_0)\, d\zeta.$$

4.2.2 Spectral Shift Function for Resolvent Comparable Operators

We let A, A_0 be self-adjoint operators in the Hilbert space $\mathcal{H}$ such that

$$(A - z)^{-1} - (A_0 - z)^{-1} \in \mathcal{L}_1(\mathcal{H}), \quad z \in \mathbb{C} \setminus \mathbb{R}.$$

Consider the Cayley transform for A and A_0 defined by

$$U = (A + i)(A - i)^{-1}, \quad U_0 = (A_0 + i)(A_0 - i)^{-1}.$$

Both U and U_0 are unitary operators and

$$U - U_0 = -2i\big[(A - i)^{-1} - (A_0 - i)^{-1}\big] \in \mathcal{L}_1(\mathcal{H}). \tag{4.22}$$

By (4.22) and Theorem 4.2.1 the spectral shift function $\eta(\cdot; U, U_0)$ exists (and defined up to an additive constant).

Therefore, one can define

$$\xi(\lambda; A, A_0) := \eta\big((\lambda + i)(\lambda - i)^{-1}; U, U_0\big), \quad \lambda \in \mathbb{R}$$

and this spectral shift function is also defined only up to an additive constant.

Recall (see e.g. [Yaf92, Section 8.1]) that for a pair of resolvent comparable operators (A, A_0), one can define the generalized perturbation determinant $\widetilde{D}_{A/A_0}(\cdot)$ by

$$\widetilde{D}_{A/A_0}(z) := \Delta_{U/U_0}\big((z+i)(z-i)^{-1}\big) = \det\Big(1+(z-i)(A-i)^{-1}B(A_0-z)^{-1}\Big), \quad a \in \mathbb{C},\ \operatorname{Im}(a) > 0.$$

The generalized perturbation determinant $\widetilde{D}_{A/A_0}(\cdot)$ is analytic in the open upper and lower complex half-planes and never vanishes on $\mathbb{C}\setminus\mathbb{R}$ [Yaf92, Section 8.1].

Theorem 4.2.1 implies the following

Theorem 4.2.4 *Let the self-adjoint operators A_0, A be such that*

$$(A - z)^{-1} - (A_0 - z)^{-1} \in \mathcal{L}_1(\mathcal{H}), \quad z \in \rho(A_0) \cap \rho(A).$$

The representation

$$\ln \widetilde{D}_{A/A_0}(z) = (z-i)\int_{\mathbb{R}} \xi(\lambda; A, A_0)(\lambda - z)^{-1}(\lambda - i)^{-1} d\lambda,$$

holds, where $\xi(\cdot\,; A, A_0)$ *is defined via*

$$\xi(\lambda; A, A_0) = \frac{1}{2\pi}\Big(\lim_{\varepsilon\downarrow 0} \mathrm{Im}\Big(\ln\big(\widetilde{D}_{A/A_0}(\lambda + i\varepsilon)\big)\Big) - \lim_{\varepsilon\downarrow 0} \mathrm{Im}\Big(\ln\big(\widetilde{D}_{A/A_0}(\lambda - i\varepsilon)\big)\Big)\Big)$$

for a.e. $\lambda \in \mathbb{R}$.

We emphasize again that in the resolvent comparable case, $\xi(\cdot; A, A_0)$ is defined only *up to an additive constant.*

Since $\eta(\cdot; U(a), U_0(a)) \in L_1(\mathbb{T})$, it follows that $\xi(\cdot; A, A_0) \in L_1\big(\mathbb{R}; (1 + \lambda^2)^{-1} d\lambda\big)$. The class of functions in Theorem 4.2.3 can be used to find the class of functions such that the Krein's trace formula holds for the pair (A, A_0). We recall that the class of functions $\mathfrak{F}_1(\mathbb{R})$ is defined in Definition 2.3.5.

Theorem 4.2.5 *Let self-adjoint operators* A_0, A *be such that*

$$(A - z)^{-1} - (A_0 - z)^{-1} \in \mathcal{L}_1(\mathcal{H}), \quad z \in \rho(A_0) \cap \rho(A)$$

and let $f \in \mathfrak{F}_1(\mathbb{R})$. *Then* $f(A) - f(A_0) \in \mathcal{L}_1(\mathcal{H})$ *and for the spectral shift function* $\xi(\cdot; A, A_0) \in L_1(\mathbb{R}; (\lambda^2 + 1)^{-1} d\lambda)$ *defined in Theorem 4.2.4 the trace formula*

$$\mathrm{tr}(f(A) - f(A_0)) = \int_{\mathbb{R}} f'(\lambda)\xi(\lambda; A, A_0) d\lambda$$

holds.

Just as in the case of a trace-class perturbation, the spectral shift function for resolvent comparable operators A_0, A possesses the following property.

Proposition 4.2.6 *Suppose that in some interval* (a_0, b_0) *the spectrum of* A_0 *is discrete and let* $\delta = (a, b)$, $a_0 < a < b < b_0$. *Then*

$$\xi(b_-; A, A_0) - \xi(a_+; A, A_0) = N_{A_0}(\delta) - N_A(\delta), \tag{4.23}$$

where $N_{A_0}(\delta)$ *(respectively,* $N_A(\delta)$*) is the sum of the multiplicities of the eigenvalues of* A_0 *(respectively,* A*) in* δ.

In the particular case of *lower semibounded operators* A_0 and A equality (4.23) allows us to *naturally fix* the additive constant in the following way. To the left of the spectra of A_0 and A, the eigenvalue counting functions $N_{A_0}(\cdot)$ and $N_A(\cdot)$ are zero. Therefore, by equality (4.23) the spectral shift function $\xi(\cdot\,; A, A_0)$ is a constant to the left of the spectra of A_0 and A, and it is custom to set this constant equal to zero,

$$\xi(\lambda; A, A_0) = 0, \quad \lambda < \inf(\sigma(A_0) \cup \sigma(A)).$$

In the general case of self-adjoint resolvent comparable operators the constant may be fixed under some additional assumptions. One way of doing this we discuss next, while a second method, which is applicable to our setting, we discuss in Sect. 4.3.

4.2.3 Invariance Principle

We now describe a particular way to introduce the spectral shift function for the pair (A, A_0) by what is usually called the *invariance principle*. This method is often used to also fix the additive constant for the resolvent comparable case.

Let Ω be an interval containing the spectra of A_0 and A, and let f be an arbitrary bounded (strictly) monotone "sufficiently" smooth function on Ω. Suppose that

$$[f(A) - f(A_0)] \in \mathcal{L}_1(\mathcal{H})$$

then, the spectral shift function $\xi(\,\cdot\,; A, A_0)$ can be defined as follows:

$$\xi(\lambda; A, A_0) = \operatorname{sgn}\big(f'(\lambda)\big)\xi(f(\lambda); f(A), f(A_0)). \tag{4.24}$$

For the function $\xi(\,\cdot\,; A, A_0)$ the Lifshitz–Krein trace formula (4.8) holds for some class of admissible functions. The latter class depends on f.

We note the following result (see [Yaf92, Sect. 8.11]):

Proposition 4.2.7 *Let* $(A - A_0) \in \mathcal{L}_1(\mathcal{H})$. *Then the spectral shift functions for the pairs* (A, A_0) *and* $(f(A), f(A_0))$ *are associated via equality* (4.24) *up to an additive, integer-valued constant.*

The assumption that the function f is monotone is crucial in the invariance principle, since it allows one to take the inverse of f in the Lipschitz-Krein trace formula. Namely, for a sufficiently nice function h we have

$$\begin{aligned} \operatorname{tr}\big(h(A) - h(A_0)\big) &= \operatorname{tr}\big((h \circ f^{-1})(f(A)) - (h \circ f^{-1})(f(A_0))\big) \\ &= \int_{\mathbb{R}} (h \circ f^{-1})'(\mu)\,\xi(\mu; f(A), f(A_0))d\mu. \end{aligned}$$

Using the substitution $\mu = f(\lambda)$, one can obtain that

$$\operatorname{tr}\big(h(A) - h(A_0)\big) = \int_{\mathbb{R}} h'(\lambda)\,\xi(f(\lambda); f(A), f(A_0))d\lambda$$
$$= \int_{\mathbb{R}} h'(\lambda)\,\xi(\lambda; A, A_0)d\lambda,$$

where the last equation follows from (4.24).

However, as discovered in [LSVZ18], the invariance principle does not hold for the operators of primary interest to us, namely the Dirac operator on $\mathbb{R}^d$, whenever $d \geq 2$.

Example 4.2.8 Let $d \geq 2$. Consider the Dirac operator $\mathcal{D} = \sum_{k=1}^{d} e_k \otimes D_k$ on $\mathbb{R}^d$ (see Sect. 7.1 for a precise definition) with electro-magnetic potential $V = 1 \otimes \varphi + \sum_{k=1}^{d} e_k \otimes a_k$ with $\varphi, a_j \in S(\mathbb{R}^d)$, $j = 1, \ldots d$. It is proved in [LSVZ18, Theorem 1.2] that if one takes a monotone real-valued function f on $\mathbb{R}$ such that $f' \in S(\mathbb{R})$ then

$$f(\mathcal{D} + V) - f(\mathcal{D}) \notin \mathcal{L}_1(\mathbb{C}^{n(d)} \otimes L_2(\mathbb{R}^d)).$$

That is the operator $f(\mathcal{D} + V) - f(\mathcal{D})$ is never trace-class in this example.

This example demonstrates that the invariance principle is not generally applicable. However, the free Dirac operator $\mathcal{D}$ and its perturbation $\mathcal{D} + V$ are d-resolvent comparable for a sufficiently good potential V (see Corollary 3.1.8), that is

$$(\mathcal{D} + V - z)^{-d} - (\mathcal{D} - z)^{-d} \in \mathcal{L}_1(\mathcal{H}), \quad z \in \mathbb{C} \setminus \mathbb{R}.$$

In [Yaf05] Yafaev presented construction of the spectral shift function for m-resolvent comparable operators. We turn to his method now.

4.2.4 *Spectral Shift Function for m-Resolvent Comparable Operators*

Suppose that A_0 and A are fixed self-adjoint operators in the Hilbert space $\mathcal{H}$, which are m-resolvent comparable in $\mathcal{L}_1(\mathcal{H})$ (see Definition 2.3.1) for some odd $m \in \mathbb{N}$. That is for all $a \in \mathbb{R} \setminus \{0\}$ we have

$$\left[(A - ai)^{-m} - (A_0 - ai)^{-m}\right] \in \mathcal{L}_1\big(\mathcal{H}\big).$$

As in Sect. 2.3 we denote by $\varphi : \mathbb{R} \to \mathbb{R}$ a bijection satisfying for some $c > 0$,

$$\varphi \in C^2(\mathbb{R}), \quad \varphi(\lambda) = \lambda^m, \ |\lambda| \geq 1, \quad \varphi'(\lambda) \geq c.$$

By Proposition 2.3.4 we have that

$$\left[(\varphi(A)-i)^{-1}-(\varphi(A_0)-i)^{-1}\right]\in\mathcal{L}_1(\mathcal{H}).$$

Therefore, by Theorem 4.2.4 there exists the (class) of spectral shift functions $\xi(\cdot;\varphi(A),\varphi(A_0))$ for the pair $(\varphi(A),\varphi(A_0))$ satisfying

$$\xi(\,\cdot\,;\varphi(A),\varphi(A_0))\in L_1\big(\mathbb{R};(\mu^2+1)^{-1}d\mu\big). \tag{4.25}$$

Hence one can introduce the (class) of spectral shift functions $\xi(\cdot;A,A_0)$ for the pair (A,A_0) by setting

$$\xi(\lambda;A,A_0):=\xi(\varphi(\lambda);\varphi(A),\varphi(A_0)),\quad \lambda\in\mathbb{R}. \tag{4.26}$$

In particular, the condition $\varphi'(\lambda)\geq c>0$ implies that the inverse function φ^{-1} is differentiable. Therefore, the simple substitution

$$\mu=\varphi(\lambda)\in\mathbb{R},\quad \lambda\in\mathbb{R}, \tag{4.27}$$

in (4.25) implies that

$$\xi(\,\cdot\,;A,A_0)\in L_1\big(\mathbb{R};(|\lambda|^{m+1}+1)^{-1}d\lambda\big).$$

Furthermore, for $f\in\mathfrak{F}_m(\mathbb{R})$, the fact that $\varphi(\lambda)=\lambda^m$ for sufficiently large (in absolute value) λ (see (2.14)) implies that $f\circ\varphi^{-1}\in\mathfrak{F}_1(\mathbb{R})$. Hence, using Theorem 4.2.4 and the change of variables (4.27), the corresponding trace formula is of the form

$$\begin{aligned}\operatorname{tr}(f(A)-f(A_0))&=\operatorname{tr}\big((f\circ\varphi^{-1})(\varphi(A))-(f\circ\varphi^{-1})(\varphi(A_0))\big)\\&=\int_{\mathbb{R}}(f\circ\varphi^{-1})'(\mu)\,\xi(\mu;\varphi(A),\varphi(A_0))d\mu\\&=\int_{\mathbb{R}}f'(\lambda)\,\xi(\varphi(\lambda);\varphi(A),\varphi(A_0))d\lambda\\&=\int_{\mathbb{R}}f'(\lambda)\,\xi(\lambda;A,A_0)d\lambda,\quad f\in\mathfrak{F}_m(\mathbb{R}),\end{aligned}$$

where the last equality follows from (4.26).

Thus, we have the following result.

Theorem 4.2.9 ([Yaf05]) *Suppose that operators A_0 and A are m-resolvent comparable with $m\in\mathbb{N}$ odd. Then there exits spectral shift function $\xi(\cdot;A,A_0)$, satisfying*

$$\xi(\,\cdot\,;A,A_0)\in L_1\big(\mathbb{R};(|\lambda|^{m+1}+1)^{-1}d\lambda\big)$$

and

$$\operatorname{tr}(f(A) - f(A_0)) = \int_{\mathbb{R}} f'(\lambda)\,\xi(\lambda; A, A_0)d\lambda, \quad f \in \mathfrak{F}_m(\mathbb{R}). \tag{4.28}$$

A combination of Theorems 3.1.7 and 4.2.9 imply the following

Corollary 4.2.10 *Assume that B is m-relative trace-class perturbation of A_0 for an odd $m \in \mathbb{N}$. Then there exists a spectral shift function $\xi(\cdot; A, A_0) \in L_1\big(\mathbb{R}; (|\lambda|^{m+1} + 1)^{-1}d\lambda\big)$ satisfying* (4.28).

The properties of the spectral shift function $\xi(\cdot, A, A_0)$ can be easily derived from the properties of the spectral shift function $\xi(\cdot; \varphi(A), \varphi(A_0))$. We note one of these properties in the following proposition.

Proposition 4.2.11 *Let A_0 and A be m-resolvent comparable with $m \in \mathbb{N}$ odd and assume that in some interval (a_0, b_0) the spectra of A_0 and A are discrete and let $\delta = (a, b)$, $a_0 < a < b < b_0$. Then,*

$$\xi(b_-; A, A_0) - \xi(a_+; A, A_0) = N_{A_0}(\delta) - N_A(\delta), \tag{4.29}$$

where $N_{A_0}(\delta)$ (respectively, $N_A(\delta)$) are the sum of the multiplicities of the eigenvalues of A_0 (respectively, A) in δ.

We note that since the spectral shift function $\xi(\cdot; \varphi(A), \varphi(A_0))$ is a priori defined only up to an additive constant, the spectral shift function $\xi(\cdot; A, A_0)$ is also defined only up to an additive constant.

As noted in Corollary 3.1.8 the Dirac operator $\mathcal{D}$ on $\mathbb{R}^d$ and its perturbation by a sufficiently good potential V satisfy

$$(\mathcal{D} + V - z)^{-d} - (\mathcal{D} - z)^{-d} \in \mathcal{L}_1(\mathcal{H}), \quad z \in \mathbb{C} \setminus \mathbb{R}.$$

The latter inclusion allows us to introduce the spectral shift function $\xi(\cdot; \mathcal{D} + V, \mathcal{D})$ in all dimensions. Recall that Theorem 4.2.9 required m to be odd, so we must take $(d+1)^{th}$ power of resolvents whenever d is even. In terms of the integrability condition for the spectral shift function we then have

$$\xi(\cdot; \mathcal{D} + V, \mathcal{D}) \in \begin{cases} L_1\big(\mathbb{R}; (|\lambda|^{d+1} + 1)^{-1}d\lambda\big), & d \text{ is odd;} \\ L_1\big(\mathbb{R}; (|\lambda|^{d+2} + 1)^{-1}d\lambda\big), & d \text{ is even.} \end{cases}$$

However, for the Dirac operator we have a stronger condition that

$$V(\mathcal{D}^2 + 1)^{-d/2+\varepsilon} \in \mathcal{L}_1(\mathcal{H}), \quad \varepsilon > 0.$$

It turns out that this stronger assumption allows us to improve the integrability condition in the even dimensional case to $\xi(\cdot; \mathcal{D} + V, \mathcal{D}) \in L_1\big(\mathbb{R}; (|\lambda|^{d+1} +$

$1)^{-1}d\lambda)$, so that it is consistent in all dimensions (whether even or odd). We present now the construction from [CGL+22].

For the remainder of this section, we assume that A_0 is a self-adjoint operator, B is a self-adjoint bounded perturbation such that for some even $m \in \mathbb{N}$ and $0 < \varepsilon < \frac{1}{2}$ we have

$$B(A_0^2 + 1)^{-m/2+\varepsilon} \in \mathcal{L}_1(\mathcal{H}).$$

Consider the function

$$\varphi(t) = t\big(1+t^2\big)^{\frac{m-1}{2}} = t\big(1+t^2\big)^{k-(1/2)}, \quad t \in \mathbb{R}$$

with k such that $m = 2k$.

Note that

$$\begin{aligned}
[\varphi(t)+i]^{-1} &= \Big[t\big(1+t^2\big)^{k-\frac{1}{2}} + i\Big]^{-1} \\
&= \frac{t\big(1+t^2\big)^{k-\frac{1}{2}}}{t^2\big(1+t^2\big)^{2k-1}+1} - i\frac{1}{t^2\big(1+t^2\big)^{2k-1}+1} \\
&= \frac{t}{\big(t^2+1\big)^{1/2}} \frac{\big(1+t^2\big)^k}{t^2\big(1+t^2\big)^{2k-1}+1} - i\frac{1}{t^2\big(1+t^2\big)^{2k-1}+1},
\end{aligned}$$

and therefore, introducing the functions

$$g(t) = \frac{t}{(t^2+1)^{1/2}}, \quad h_1(t) = \frac{\big(1+t^2\big)^k}{t^2\big(1+t^2\big)^{2k-1}+1}, \quad h_2(t) = \frac{1}{t^2\big(1+t^2\big)^{2k-1}+1}, \quad t \in \mathbb{R},$$

we have

$$(\varphi(t)+i)^{-1} = g(t)h_1(t) - ih_2(t).$$

Thus,

$$\begin{aligned}
&\big[(\varphi(A)+i)^{-1} - (\varphi(A_0)+i)^{-1}\big] \\
&= g(A)h_1(A) - g(A_0)h_1(A_0) - i[h_2(A) - h_2(A_0)] \\
&= [g(A) - g(A_0)]h_1(A_0) + g(A)[h_1(A) - h_1(A_0)] - i[h_2(A) - h_2(A_0)].
\end{aligned}$$

As shown in [CGL+22, Lemma 8.10],

$$[g(A) - g(A_0)]h_1(A_0) \in \mathcal{L}_1(\mathcal{H}).$$

At the same time, $h_i(t) = f_i((1+t^2)^{-k})$, for some $f_i \in C^{1,\frac{1}{2}}([0,1])$, $i = 1, 2$ (see [CGL+22, Lemma 8.9]) and so

$$h_j(A) - h_j(A_0) = f_j\Big(\big(A^2+1\big)^{-k}\Big) - f_j\Big(\big(A_0^2+1\big)^{-k}\Big), \quad j = 1, 2.$$

Lemma 3.1.9 and Theorem 2.2.10 imply that

$$h_j(A) - h_j(A_0) \in \mathcal{L}_1(\mathcal{H}), \quad j = 1, 2.$$

Therefore,

$$\big[\varphi(A) + i)^{-1} - (\varphi(A_0) + i)^{-1}\big] \in \mathcal{L}_1(\mathcal{H}).$$

By Theorem 4.2.4 the spectral shift function $\xi(\,\cdot\,; \varphi(A), \varphi(A_0))$ exists. As φ is a strictly increasing function on $\mathbb{R}$, using the invariance invariance principle for the spectral shift function we can introduce $\xi(\,\cdot\,; A, A_0)$ by setting

$$\xi(\lambda; A, A_0) = \xi(\varphi(\lambda); \varphi(A), \varphi(A_0)) \text{ for a.e. } \lambda \in \mathbb{R}.$$

The choice of φ and integrability properties of $\xi(\,\cdot\,; \varphi(A), \varphi(A_0))$ will imply an appropriate integrability condition for $\xi(\,\cdot\,; A, A_0)$.

Thus, we obtain the following theorem, which improves the integrability condition in Theorem 4.2.9 for even $m \in \mathbb{N}$.

Theorem 4.2.12 *Assume that A_0 is a self-adjoint operator, B is a self-adjoint bounded perturbation such that for some $m \in \mathbb{N}$ and $0 < \varepsilon < \frac{1}{2}$ we have*

$$B(A_0^2 + 1)^{-m/2+\varepsilon} \in \mathcal{L}_1(\mathcal{H}).$$

For any $f \in \mathfrak{F}_m(\mathbb{R})$ one has

$$[f(A) - f(A_0)] \in \mathcal{L}_1(\mathcal{H}),$$

and there exists a function

$$\xi(\,\cdot\,; A, A_0) \in L_1\big(\mathbb{R}; (1+|\lambda|)^{-m-1}\, d\lambda\big)$$

such that the following trace formula holds,

$$\operatorname{tr}(f(A) - f(A_0)) = \int_{\mathbb{R}} f'(\lambda)\xi(\lambda; A, A_0)\, d\lambda, \quad f \in \mathfrak{F}_m(\mathbb{R}).$$

Remark 4.2.13 Assume that A_0, A and B are as in Theorem 4.2.12 and let m be even. On one hand, Theorem 4.2.12 guarantees that there exists spectral shift function $\xi_1(\cdot; A, A_0) \in L_1\big(\mathbb{R}; (1+|\lambda|)^{-m-1}\, d\lambda\big)$ satisfying (6.19). On the other hand, since B is $p+1$-relative perturbation on A_0, Corollary 4.2.10 implies that there exists spectral shift function $\xi_2(\cdot; A, A_0) \in L_1\big(\mathbb{R}; (1+|\lambda|)^{-m-2}\, d\lambda\big)$ satisfying (4.28). In particular, for any $f \in C_0^\infty(\mathbb{R})$, we have that

$$\int_{\mathbb{R}} f'(\lambda)\xi_1(\lambda; A, A_0)d\lambda = \int_{\mathbb{R}} f'(\lambda)\xi_2(\lambda; A, A_0)d\lambda.$$

Referring to the Du Bois-Raymond Lemma (see, e.g., [LL01, Theorem 6.11]), the functions $\xi_1(\cdot; A, A_0)$ and $\xi_2(\cdot; A, A_0)$ differ a.e. at most by a constant. ◇

4.3 Continuity of the Spectral Shift Function with Respect to the Operator Parameter

Here we address continuity of $\xi(\cdot; A, A_0)$ with respect to the operator parameter A (assuming that A_0 is fixed).

Let T be some fixed self-adjoint operator on $\mathcal{H}$. Denote by $\Gamma(T)$ the space of all self-adjoint operators, which are resolvent comparable with T, that is A on $\mathcal{H}$ such that

$$(T-z)^{-1} - (A-z)^{-1} \in \mathcal{L}_1(\mathcal{H})$$

for some $z \in \mathbb{C} \setminus \mathbb{R}$. For every $z \in \mathbb{C}$, $\mathrm{Im}(z) > 0$, one defines a metric on $\Gamma(T)$ by setting

$$d_z(A_1, A_2) = 2\mathrm{Im}(z)\big\|(A_2-z)^{-1} - (A_1-z)^{-1}\big\|_1.$$

By a standard resolvent identity, the metrics d_{z_1} and d_{z_2} are equivalent for different values of z_1 and z_2 in $\mathbb{C} \setminus \mathbb{R}$.

Proposition 4.3.1 ([Yaf92, Lemma 8.7.5]) *Let A_0, A, and B_1 denote self-adjoint operators in $\mathcal{H}$ with $B_0, B_1 \in \Gamma(A_0)$, and let $\{B(s)\}_s \subset \Gamma(A)$ be a continuous (with respect to s) path from B_0 to B_1 in $\Gamma(A)$. Assume also that the spectral shift function $\xi_0(\cdot; B_0, A_0)$ is fixed representative from the class $\xi(\cdot; B_0, A_0)$. Then, there exists a unique representative $\xi(\cdot; B(s), A_0)$, continuous in s with respect to the norm in $L_1(\mathbb{R}, (\lambda^2+1)^{-1}d\lambda)$, such that*

$$\xi(\cdot; B(0), A_0) = \xi_0(\cdot; B_0, A_0).$$

In the rest of this section we discuss the analogue of Proposition 4.3.1 for m-resolvent comparable operators.

Definition 4.3.2 Let T be self-adjoint in $\mathcal{H}$ and $m \in \mathbb{N}$ odd. Then $\Gamma_m(T)$ denotes the set of all self-adjoint operators A in $\mathcal{H}$ for which the inclusion

$$\left[(A - ai)^{-m} - (T - ai)^{-m}\right] \in \mathcal{L}_1(\mathcal{H}), \quad a \in \mathbb{R}\backslash\{0\},$$

holds. In particular, $\Gamma_1(T) = \Gamma(T)$.

We note that for each $m \in \mathbb{N}$, $\Gamma_m(T)$ can be equipped with the family $\mathcal{D} = \{d_{m,a}\}_{a\in\mathbb{R}\backslash\{0\}}$ of pseudometrics (see [Dug66, Definition IX.10.1] for a precise definition) defined by

$$d_{m,a}(A_1, A_2) = \left\|(A_2 - ai)^{-m} - (A_1 - ai)^{-m}\right\|_1, \quad A_1, A_2 \in \Gamma_m(T). \tag{4.30}$$

For each fixed $\varepsilon > 0$, $a \in \mathbb{R}\backslash\{0\}$, and $A \in \Gamma_m(T)$, define

$$B(A; d_{m,a}, \varepsilon) = \{A' \in \Gamma_m(T) \,|\, d_{m,a}(A, A') < \varepsilon\},$$

to be the ε-ball centered at A with respect to the pseudometric $d_{m,a}$.

Definition 4.3.3 $\mathcal{T}_m(T)$ is the topology on $\Gamma_m(T)$ with the subbasis

$$\mathfrak{B}_m(T) = \{B(A; d_{m,a}, \varepsilon) \,|\, A \in \Gamma_m(T),\ a \in \mathbb{R}\backslash\{0\},\ \varepsilon > 0\}.$$

That is, $\mathcal{T}_m(T)$ is the smallest topology on $\Gamma_m(T)$ which contains $\mathfrak{B}_m(T)$.

If φ is the function introduced in (2.14), then Proposition 2.3.4 implies that $\{\varphi(B(s))\}_{s\in[0,1]} \in \Gamma(\varphi(A_0))$ is continuous in $\Gamma(\varphi(A_0))$ with respect to the metric $d_z(\cdot,\cdot)$ for any path $\{B(s)\}_{s\in[0,1]} \in \Gamma_m(A_0)$ which is continuous with respect to the topology $\mathcal{T}_m(A_0)$. If we now assume that the spectral shift function $\xi_0(\cdot; B_0, A_0) = \xi_0(\varphi(\cdot); \varphi(B_0), \varphi(A_0))$ is fixed, then Proposition 4.3.1 implies that there exists a unique spectral shift function $\xi(\,\cdot\,; \varphi(B(s)), \varphi(A_0)) \in L_1(\mathbb{R}; (\lambda^2 + 1)^{-1}d\lambda)$, depending continuously on $s \in [0, 1]$ in the $L_1(\mathbb{R}; (\lambda^2 + 1)^{-1}d\lambda)$-norm and such that

$$\xi(\,\cdot\,; B(0), A_0) = \xi(\varphi(\cdot), \varphi(B(0)), \varphi(A_0)) = \xi_0(\varphi(\cdot), \varphi(B_0), \varphi(A_0)) = \xi_0(\cdot, B_0, A_0).$$

Some elementary estimates and continuity of the spectral shift function $\xi(\,\cdot\,; \varphi(B(s)), \varphi(A_0))$ (with respect to s) in the $L_1(\mathbb{R}; (\lambda^2 + 1)^{-1}d\lambda)$-norm, then imply that there exists a unique spectral shift function $\xi(\cdot; B(s), A_0)$, continuous in s with respect to the norm in $L_1(\mathbb{R}, (|\lambda|^{p+1} + 1)^{-1}d\lambda)$, such that

$$\xi(\cdot; B(0), A_0) = \xi_0(\cdot; B_0, A_0).$$

Namely, we have the following theorem, which will allow us to fix the undefined additive constant in the spectral shift function in Theorem 6.4.11.

Theorem 4.3.4 ([CGL⁺16a]) *Let A_0, B_0, and B_1 denote self-adjoint operators in $\mathcal{H}$ with $B_0, B_1 \in \Gamma_m(A_0)$, and let $\{B(s)\}_s \subset \Gamma(A)$ be a continuous (with respect to s) path from B_0 to B_1 in the topology $\mathcal{T}_m(A)$. Assume also that the spectral shift function $\xi_0(\cdot; B_0, A_0)$, defined by (4.26), is a fixed representative from the class $\xi(\cdot; B_0, A_0)$. Then, there exists a unique representative $\xi(\cdot; B(s), A_0)$, continuous in s with respect to the norm in $L_1(\mathbb{R}, (|\lambda|^{m+1}+1)^{-1}d\lambda)$, such that*

$$\xi(\cdot; B(0), A_0) = \xi_0(\cdot; B_0, A_0).$$

Theorem 4.3.4 fixes the additive constant in the spectral shift function, in the case when the perturbation is given by p-relative trace-class perturbation, as follows.

Let A_0 be a self-adjoint operator and let

$$P_n = \chi_{[-n,n]}(A_0), \quad n \in \mathbb{N}$$

and let B be an p-relative trace-class perturbation on A_0, $p \in \mathbb{N}$. By Theorems 4.2.9 and 3.1.7 we can define the spectral shift function $\xi(\cdot; A_0 + B, A_0)$ such that

$$\xi(\cdot; A_0 + B, A_0) \in L_1(\mathbb{R}; (|\lambda|^{p_0+1}+1)^{-1}d\lambda), \quad p_0 = 2\lfloor \frac{p}{2} \rfloor + 1.$$

Introducing the path $\{B(s)\}_{s\in[0,1]}$, by setting

$$B(s) = A_0 + \hat{P}_s B \hat{P}_s, \quad \mathrm{dom}(B(s)) = \mathrm{dom}(A_0), \quad s \in [0,1],$$
$$\hat{P}_s = \chi_{[-\frac{1}{1-s}, \frac{1}{1-s}]}(A_0), s \in [0,1), \quad \hat{P}_1 = 1,$$

joining $B(0) = A_0 + P_1 B P_1$ and $B(1) = A_0 + B$, and using Theorem 3.1.7 we have that the family $B(s)$ depends continuously on $s \in [0,1]$ with respect to the family of pseudometric $d_{p_0,a}(\cdot,\cdot)$ defined by (4.30). Furthermore, since $\hat{P}_0 B \hat{P}_0 = P_1 B P_0 \in \mathcal{L}_1(\mathcal{H})$, it follows that the spectral shift function $\xi(\cdot; A_0 + \hat{P}_0 B \hat{P}_0, A_0) \in L_1(\mathbb{R})$. Therefore, Theorem 4.3.4 implies that there exists a unique spectral shift function $\xi(\cdot; B(s), A_0)$ for the pair $(B(s), A_0)$ depending continuously on $s \in [0,1]$ in the space $L_1\big(\mathbb{R}; (|\lambda|^{p_0+1}+1)^{-1}d\lambda\big)$, satisfying $\xi(\cdot; B(0), A_0) = \xi(\cdot; A_0 + P_1 B P_1, A_0)$. For the specific values $s = \frac{n-1}{n}$ we then have

$$\xi(\cdot; A_0 + B, A_0) = \lim_{n\to\infty} \xi(\cdot; B(\frac{n-1}{n}), A_0) = \lim_{n\to\infty} \xi(\cdot; A_0 + P_n B P_n, A_0)$$

in $L_1\big(\mathbb{R}; (|\lambda|^{p_0+1}+1)^{-1}d\lambda\big)$.

Thus, we have the following result:

Theorem 4.3.5 *Suppose that A_0 is a self-adjoint operator on $\mathcal{H}$ and let $B \in \mathcal{L}(\mathcal{H})$ be a p-relative trace-class perturbation of A_0, $p \in \mathbb{N}$. Then there exists unique* *spectral shift function* $\xi(\cdot; A_0 + B, A_0)$ *such that*

$$\xi(\cdot; A_0 + B, A_0) = \lim_{n\to\infty} \xi(\cdot; A_0 + P_n B P_n, A_0) \tag{4.31}$$

in $L_1(\mathbb{R}; (|\lambda|^{p_0+1} + 1)^{-1}d\lambda)$, $p_0 = 2\lfloor \frac{p}{2} \rfloor + 1$.

We conclude with an elementary consequence of Theorem 4.3.5.

Corollary 4.3.6 *Let* A_0, B *and* P_n *be as in Theorem 4.3.5. For* $h \in L_\infty(\mathbb{R})$ *such that* $\sup_{\lambda\in\mathbb{R}} \left|(|\lambda|^{p_0+1} + 1)h(\lambda)\right| < \infty$ *we have*

$$\lim_{n\to\infty} \int_{\mathbb{R}} \xi(\lambda; A_0 + P_n B P_n, A_0)h(\lambda)d\lambda = \int_{\mathbb{R}} \xi(\lambda; A_0 + B, A_0)h(\lambda)d\lambda.$$

4.4 Representation of the Spectral Shift Function via a Regularised Perturbation Determinant

In this section we establish the representation of $\xi(\,\cdot\,; A, A_0)$ for m-resolvent comparable operators in terms of regularized Fredholm determinants.

We start with the definition of regularised (Fredholm) determinant (see e.g. [Yaf92]).

The definition of thc Fredholm determinant given in (4.2) is generally not meaningful if $T \in \mathcal{L}_p(\mathcal{H})$ with $p > 1$. To give meaning to the determinant in this case, suitable modifications are needed. For $p \in \mathbb{N}\backslash\{1\}$, one introduces the function $R_p : \mathcal{L}_p(\mathcal{H}) \to \mathcal{L}_1(\mathcal{H})$ by

$$R_p(T) = (1 + T)e^{\sum_{j=1}^{p-1}(-1)^j j^{-1}T^j} - 1, \quad T \in \mathcal{L}_p(\mathcal{H}).$$

Then the *regularized (or modified) Fredholm determinant* is defined by

$$\det_p(1 + T) = \det(1 + R_p(T)), \quad T \in \mathcal{L}_p(\mathcal{H}), \quad p \in \mathbb{N}\backslash\{1\}.$$

The Fredholm determinant $\det_p(\,\cdot\,)$ retains many of the properties of the ordinary Fredholm determinant. For example, $\det_p(1 + A)$ is continuous with respect to $T \in \mathcal{L}_p(\mathcal{H})$: if $T \in \mathcal{L}_p(\mathcal{H})$, $\{T_n\}_{n=1}^\infty \subset \mathcal{L}_p(\mathcal{H})$, and $\lim_{n\to\infty} \|T_n - T\|_{\mathcal{L}_p(\mathcal{H})} = 0$, then

$$\lim_{n\to\infty} \det_p(1 + T_n) = \det_p(1 + T).$$

In addition, if $\Omega \subseteq \mathbb{C}$ is open and $T : \Omega \to \mathcal{L}_p(\mathcal{H})$ is analytic in Ω, then the function $\det_p(1 + T(\cdot))$ is analytic in Ω and

$$\frac{d}{dz} \ln\big(\det_p(1 + T(z))\big) = (-1)^{p-1} \operatorname{tr}\big((1 + T(z))^{-1} T^{p-1}(z) T'(z)\big).$$

The importance of the regularized determinant stems from the fact that for $T \in \mathcal{L}_p(\mathcal{H})$, the operator $1 + T$ is boundedly invertible (i.e., $-1 \in \rho(T)$) if and only if $\det_p(1 + T) \neq 0$. Equivalently, $-1 \in \sigma(A)$ if and only if $\det_p(1 + T) = 0$.

For the discussion in this section we need:

Hypothesis 4.4.1 *Let A_0 and A be self-adjoint operators in $\mathcal{H}$, such that $B = A - A_0$ is a p-relative trace-class perturbation on A_0, $p \in \mathbb{N}$. If p is even assume, in addition, that for some $0 < \varepsilon < 1/2$,*

$$B\big(A_0^2 + 1\big)^{-(p/2)-\varepsilon} \in \mathcal{L}_1(\mathcal{H}).$$

Then, as A_0 and B satisfy the assumption of Theorem 4.2.12, it follows that there exists spectral shift function $\xi(\cdot; A, A_0) \in L_1\big(\mathbb{R}; (1 + |\lambda|)^{-p-1} d\lambda\big)$.

Remark 4.4.2 In this section we aim to establish an a.e. pointwise formula for the spectral shift function. For this reason, we work with the spectral shift functions $\xi(\cdot; A, A_0)$ only defined up to an additive constant and do not fix this constant in any way. ◇

By (3.2) the (non-symmetrized) Birman–Schwinger-type operator

$$B(z) := (A - A_0)(A_0 - z)^{-1}, \quad z \in \mathbb{C}\backslash\mathbb{R} \tag{4.32}$$

is an $\mathcal{L}_{p+1}(\mathcal{H})$-valued function. The resolvent decomposition then also implies that the map $z \mapsto B(z)$ is a $\mathcal{L}_{p+1}(\mathcal{H})$-valued analytic function. In particular, we can define the regularised perturbation determinant

$$\det_{p+1}(1 + (A - A_0)(A_0 - z)^{-1}),$$

which is an analytic function in $\mathbb{C}_\pm$. Similarly to (4.3), we have that $\det_{p+1}(1 + (A - A_0)(A_0 - z)^{-1}) \neq 0$ for all $z \in \mathbb{C} \setminus \mathbb{R}$. Hence, we can define the analytic function

$$z \mapsto \ln\big(\det_{p+1}\big(1 + (A - A_0)(A_0 - z)^{-1}\big)\big), \quad z \in \mathbb{C}_\pm. \tag{4.33}$$

To correlate the function $\ln\big(\det_{p+1}\big(1 + (A - A_0)(A_0 - z)^{-1}\big)\big)$ with the spectral shift function for the pair (A, A_0), we need to introduce an auxiliary function. With

$B(\cdot)$ defined in (4.32) one can show that

$$z \mapsto \operatorname{tr}\left(\frac{d^{m-1}}{dz^{m-1}}(A_0 - z)^{-1}B(z)^{m-j}\right), \quad z \in \mathbb{C}\backslash\mathbb{R},\ j \in \mathbb{N}_0,\ 0 \leq j \leq m-1,$$

defines an analytic function on $\mathbb{C}_{\pm}$.We define the auxiliary analytic function $G_{A,A_0}(\cdot)$ in $\mathbb{C}_{\pm}$ such that

$$\frac{d^p}{dz^p}G_{A,A_0}(z) = \operatorname{tr}\left(\frac{d^{p-1}}{dz^{p-1}}\sum_{j=0}^{p-1}(-1)^{p-j}(A_0 - z)^{-1}B(z)^{p-j}\right), \quad z \in \mathbb{C}\backslash\mathbb{R}. \tag{4.34}$$

The following lemma is the main result we need.

Lemma 4.4.3 ([CGL⁺22, Lemma 9.7]) *Assume Hypothesis 4.4.1. There exist polynomials $P_{\pm,p-1}$ of degree less than or equal to $p-1$ such that for all $z \in \mathbb{C}_{\pm}$ the representation*

$$\ln\left(\det_{p+1}\left(1 + (A - A_0)(A_0 - z)^{-1}\right)\right) = (z-i)^p \int_{\mathbb{R}} \frac{\xi(\lambda; A, A_0)d\lambda}{(\lambda - i)^p}\frac{1}{\lambda - z}$$
$$+G_{A,A_0}(z) + P_{\pm,p-1}(z) \tag{4.35}$$

holds. Here, G_{A,A_0} denotes the analytic function in $\mathbb{C}\backslash\mathbb{R}$ introduced in (4.34).

Proof To shorten notation we introduce

$$F_{A,A_0}(z) = \ln\left(\det_{p+1}\left(1 + (A - A_0)(A_0 - z)^{-1}\right)\right), \quad z \in \mathbb{C} \setminus \mathbb{R}.$$

One recalls the $\mathcal{L}_{p+1}(\mathcal{H})$-valued analytic function $B(\cdot)$ defined in (4.32). By the second resolvent identity, we have

$$\begin{aligned}\left(1 + B(z)\right)^{-1} &= \left(1 + (A - A_0)(A_0 - z)^{-1}\right)^{-1} \\ &= \left((A - z)(A_0 - z)^{-1}\right)^{-1} \\ &= (A_0 - z)(A - z)^{-1}, \quad z \in \mathbb{C}\backslash\mathbb{R}.\end{aligned}$$

In addition,

$$B'(z) = B(z)(A_0 - z)^{-1}, \quad z \in \mathbb{C}\backslash\mathbb{R}.$$

Since

$$\frac{d}{dz}\ln\left(\det_{p+1}(1 + A(z))\right) = (-1)^p \operatorname{tr}\left((1 + A(z))^{-1}A^p(z)A'(z)\right).$$

we have

$$\frac{d}{dz}F_{A,A_0}(z) = \frac{d}{dz}\ln\big(\det_{p+1}(1+B(z))\big)$$
$$= (-1)^p \operatorname{tr}\big((A_0-z)(A-z)^{-1}B(z)^{p+1}(A_0-z)^{-1}\big),$$
$$z \in \mathbb{C}\backslash\mathbb{R}.$$

The cyclicity of the trace and the second resolvent identity imply that

$$\frac{d}{dz}F_{A,A_0}(z) = (-1)^p \operatorname{tr}\big((A-z)^{-1}B(z)^{p+1}\big)$$
$$= (-1)^p \operatorname{tr}\big((A-z)^{-1}(A-A_0)(A_0-z)^{-1}B(z)^p\big)$$
$$= (-1)^p \operatorname{tr}\big(\big((A_0-z)^{-1}-(A-z)^{-1}\big)B(z)^p\big).$$

Using repeatedly the second resolvent identity we infer that

$$\frac{d}{dz}F_{A,A_0}(z) \tag{4.36}$$
$$= \operatorname{tr}\left((A_0-z)^{-1}-(A-z)^{-1}+(A_0-z)^{-1}\sum_{j=0}^{p-1}(-1)^{p-j}B(z)^{p-j}\right),$$
$$\in \mathbb{C}\backslash\mathbb{R}.$$

Differentiating (4.36) $p-1$ times,

$$\frac{d^p}{dz^p}F_{A,A_0}(z) = \operatorname{tr}\left((p-1)!\big((A_0-z)^{-p}-(A-z)^{-p}\big)\right.$$
$$\left.+\frac{d^{p-1}}{dz^{p-1}}(A_0-z)^{-1}\sum_{j=0}^{p-1}(-1)^{p-j}B(z)^{p-j}\right)$$
$$= (p-1)!\operatorname{tr}\left((A_0-z)^{-p}-(A-z)^{-p}\right)$$
$$+\frac{d^p}{dz^p}G_{A,A_0}(z), \quad z \in \mathbb{C}\backslash\mathbb{R}.$$

Theorem 4.2.12 implies that

$$\frac{d^p}{dz^p}F_{A,A_0}(z) = p!\int_{\mathbb{R}}\frac{\xi(\lambda;A,A_0)\,d\lambda}{(\lambda-z)^{p+1}}+\frac{d^p}{dz^p}G_{A,A_0}(z), \quad z \in \mathbb{C}\backslash\mathbb{R}.$$

Since

$$\frac{p!}{(\lambda - z)^{p+1}} = \frac{d^p}{dz^p}\left(\frac{(z-i)^p}{(\lambda - z)(\lambda - i)^p}\right), \quad z \in \mathbb{C}\backslash\mathbb{R},\ \lambda \in \mathbb{R}.$$

we conclude that

$$\frac{d^p}{dz^p} F_{A,A_0}(z) = \frac{d^p}{dz^p} \int_{\mathbb{R}} \frac{\xi(\lambda; A, A_0) d\lambda\, (z-i)^p}{(\lambda - i)^p (\lambda - z)} + \frac{d^p}{dz^p} G_{A,A_0}(z), \quad z \in \mathbb{C}\backslash\mathbb{R},$$

completing the proof. □

Introducing now

$$A_\pm(z) := \ln \det_{p+1}(I + (A - A_0)(A_0 - z)^{-1}) - G_{A,A_0}(z) - P_{\pm,p-1}(z), \quad z \in \mathbb{C}_\pm,$$

Lemma 4.4.3 can be reformulated as

$$\frac{A_\pm(z)}{(z-i)^p} = \int_{\mathbb{R}} \frac{\xi(\lambda; A, A_0) d\lambda}{(\lambda - z)(\lambda - i)^p}, \quad z \in \mathbb{C}_\pm,\ z \neq i.$$

Since the spectral shift function $\xi(\cdot\,; A, A_0)$ satisfies

$$\xi(\cdot\,; A, A_0)(\cdot - i)^{-p} \in L_1\big(\mathbb{R}; (1 + |\lambda|)^{-1}\, d\lambda\big),$$

an application of Privalov's theorem implies the following

Theorem 4.4.4 *Assume Hypothesis 4.4.1 and let G_{A,A_0} denote analytic functions in $\mathbb{C}\backslash\mathbb{R}$ satisfying* (4.34). *If* $\ln \det_{p+1}(I + (A - A_0)(A_0 - \cdot)^{-1})$ *and G_{A,A_0} have normal boundary values on $\mathbb{R}$, then for a.e. $\lambda \in \mathbb{R}$,*

$$\begin{aligned} \xi(\lambda; A, A_0) &= \pi^{-1} \mathrm{Im} \ln \det_{p+1}(1 + (A - A_0)(A_0 - (\lambda + i0))^{-1}) \\ &\quad -\pi^{-1} \mathrm{Im}(G_{A,A_0}(\lambda + i0)) + P_{p-1}(\lambda), \end{aligned} \tag{4.37}$$

where P_{p-1} is a polynomial of degree less than or equal to $p - 1$.

4.5 Spectral Shift Functions for the Pairs (A_+, A_-), $(\boldsymbol{H}_2, \boldsymbol{H}_1)$

Throughout this section we assume Hypothesis 3.2.5. We start with the pair $(\boldsymbol{H}_2, \boldsymbol{H}_1)$. By Theorem 3.2.6 we have $(\boldsymbol{H}_2 - z)^{-m} - (\boldsymbol{H}_1 - z)^{-m} \in \mathcal{L}_1\big(L_2(\mathbb{R}; \mathcal{H})\big)$, for all $z \in \mathbb{C} \setminus [0, \infty)$. Since the operators $\boldsymbol{H}_2$, $\boldsymbol{H}_1$ are non-negative, the function $\lambda \to (\lambda - z)^{-m}$ is monotone on the spectra of $\boldsymbol{H}_j$, $j = 1, 2$, for any $z < 0$. Hence, by the invariance principle for the spectral shift function (see Sect. 4.2.3) there exists a spectral shift function $\xi(\,\cdot\,; \boldsymbol{H}_2, \boldsymbol{H}_1)$ for the pair $(\boldsymbol{H}_2, \boldsymbol{H}_1)$ that satisfies

$$\xi(\,\cdot\,; \boldsymbol{H}_2, \boldsymbol{H}_1) \in L_1\big(\mathbb{R}; (|\lambda|^{m+1} + 1)^{-1} d\lambda\big).$$

Since $\boldsymbol{H}_j \geq 0$, $j = 1, 2$, $\xi(\,\cdot\,; \boldsymbol{H}_2, \boldsymbol{H}_1)$ may be specified uniquely by requiring that

$$\xi(\lambda; \boldsymbol{H}_2, \boldsymbol{H}_1) = 0, \quad \lambda < 0. \tag{4.38}$$

In addition, the trace formula

$$\operatorname{tr}\big(f(\boldsymbol{H}_2) - f(\boldsymbol{H})\big) = \int_{[0,\infty)} f'(\lambda)\xi(\lambda; \boldsymbol{H}_2, \boldsymbol{H}_1)\, d\lambda, \quad f \in \mathfrak{F}_m(\mathbb{R}) \tag{4.39}$$

holds.

We introduce now the spectral shift function for the pair (A_+, A_-). Let, as before,

$$p_0 = 2\lfloor \frac{p}{2} \rfloor + 1.$$

By Theorem 3.1.7 we have that $(A_+ - z)^{-p_0} - (A_- - z)^{-p_0} \in \mathcal{L}_1(\mathcal{H})$, and therefore, by Theorem 4.2.9 there exists a spectral shift function

$$\xi(\cdot; A_+, A_-) \in L_1(\mathbb{R}; (1 + |\lambda|)^{-p_0-1}\, d\lambda) \tag{4.40}$$

such that

$$\operatorname{tr}\, (f(A_+) - f(A_-)) = \int_{\mathbb{R}} f'(\lambda)\xi(\lambda; A_+, A_-)\, d\lambda, \quad f \in \mathfrak{F}_{p_0}(\mathbb{R}). \tag{4.41}$$

Furthermore, since B_+ is a p-relative trace-class perturbation, Theorem 4.3.5 implies that there exists a unique spectral shift function $\xi(\cdot; A_+, A_-)$ such that

$$\xi(\cdot; A_+, A_-) = \lim_{n\to\infty} \xi(\cdot; A_{+,n}, A_-) \tag{4.42}$$

in $L_1(\mathbb{R}; (|\lambda|^{p_0+1} + 1)^{-1} d\lambda)$.

Since every $\xi(\,\cdot\,; A_{+,n}, A_-)$, $n \in \mathbb{N}$, is uniquely defined, Theorem 4.31 implies that we can fix uniquely the spectral shift function $\xi(\,\cdot\,; A_+, A_-)$ satisfying Eq. (4.42). We adopt this method of fixing the spectral shift function $\xi(\cdot; A_+, A_-)$ for the remainder of these notes.

Chapter 5
Spectral Flow

In the present chapter we introduce spectral flow analytically and then explain the history of formulas for spectral flow. We then discuss the connection between the spectral shift function $\xi(\cdot; A_+, A_-)$ and the spectral flow for the path $\{A(t)\}_{t\in\mathbb{R}}$. We begin, in Sect. 5.1, with the analytic definition of spectral flow due to Phillips. We recapitulate the origin of formulas for spectral flow in the 'variation of eta formula'. We then describe some of the analytic formulas that were proved subsequently for spectral flow. Our formulation of these analytic formulas relies on the results in [CPS09] (see Theorem 5.1.3). This more recent formulation provides the right viewpoint for our approach, since it applies to the path of reduced operators $\{A_n(t)\}_{t\in\mathbb{R}}$.

We follow this introductory material with Sect. 5.2 where, under the additional assumption that the asymptotes $A_\pm$ are Fredholm which guarantees that the spectral flow $\mathrm{sf}(\{A(t)\}_{t\in\mathbb{R}})$ for the family $\{A(t)\}_{t\in\mathbb{R}}$ exists, we show that:

$$\begin{aligned}&\frac{1}{2}\big(\xi(0_+; A_+, A_-) + \xi(0_-; A_+, A_-)\big)\\&\quad = \mathrm{sf}\{A(t)\}_{t=-\infty}^{\infty} - \frac{1}{2}[\dim(\ker(A_+)) - \dim(\ker(A_-))].\end{aligned} \tag{5.1}$$

This result generalises [ACS07] by allowing some essential spectra for the operators $A_\pm$.

Finally, in Sect. 5.3 we remove the assumption that the asymptotes $A_\pm$ are Fredholm, in which case the classical definition of the spectral flow does not apply. We propose there a definition of a 'generalised spectral flow' for the situation when the operators in the path are not Fredholm. Most importantly we prove that the notion of generalised spectral flow is independent of the path chosen to join the asymptotes $A_\pm$ subject of course to some simple constraints. The proof relies on establishing that the integrand in the formula defining generalised spectral flow is an exact one form on a Banach manifold (an affine space) of perturbations of a fixed self-adjoint operator A_-.

A. Carey, G. Levitina, *Index Theory Beyond the Fredholm Case*, Lecture Notes in Mathematics 2323, https://doi.org/10.1007/978-3-031-19436-8_5

5.1 Phillips' Definition of Spectral Flow and Analytic Formulas

In order to understand how we approach spectral flow in these notes we need to recapitulate analytic methods from the last 25 years. It was J. Phillips [Phi96] who introduced an analytic definition of spectral flow along a norm continuous path of bounded Fredholm operators $\{F_t : t \in [0, 1]\}$. His definition is more useful than the original approach of Atiyah-Patodi-Singer. We summarise this analytic viewpoint now.

Phillips' approach [Phi96] starts with the following observation. Let χ denote the characteristic function of the interval $[0, \infty)$. If $\{F_t\}$ is any norm continuous path of self-adjoint Fredholm operators then $\{\chi(F_t)\}$ is, in general, a discontinuous path of projections whose discontinuities arise precisely because of spectral flow. For example, if $t_1 < t_2$ are neighbouring path parameters and the projections $P_i = \chi(F_{t_i})$ commute, then the spectral flow from t_1 to t_2 should be trace$(P_2 - P_1P_2)$ minus trace$(P_1 - P_1P_2)$ (= amount of nonnegative spectrum gained minus amount of nonnegative spectrum lost). However, this is clearly the index of the operator $P_1P_2 : P_2(\mathcal{H}) \to P_1(\mathcal{H})$.

If these projections do not commute then one can still make sense of this index provided $\pi(P_1) = \pi(P_2)$ in the Calkin algebra. This notion was originally called the *essential codimension* in the type I_∞ case and denoted by $ec(P_1, P_2)$. However this terminology has subsequently fallen into disuse. Phillips' new ingredient is the fact that the operator $P_1P_2 : P_2(\mathcal{H}) \to P_1(\mathcal{H})$ is a Fredholm operator if and only if $\|\pi(P_1) - \pi(P_2)\| < 1$. While Phillips only proved the sufficient condition a complete account of this basic lemma was given in [BCP+06] including the fact that everything transports over to the most general possible case of spectral flow in a general semifinite von Neumann algebra.

Continuing now with Phillips' definition of spectral flow, we denote by π the projection onto the Calkin algebra $\mathcal{L}(\mathcal{H})/\mathcal{K}(\mathcal{H})$. Then one may show that $\pi(\chi(F_t)) = \chi(\pi(F_t))$. Since the spectra of $\pi(F_t)$ are bounded away from 0, this latter path is continuous. By compactness we can choose a partition $0 = t_0 < t_1 < \ldots < t_k = b$ so that for each $i = 1, 2, \ldots, k$

$$\|\pi(\chi(F_t)) - \pi(\chi(F_s))\| < 1/2 \quad \text{for all } t, s \in [t_{i-1}, t_i].$$

Letting $P_i = \chi(F_{t_i})$ for $i = 0, 1, \ldots, k$, then by the previous inequality (see [6]) the operator $P_{i-1}P_i : P_i(\mathcal{H}) \to P_{i-1}(\mathcal{H})$ is Fredholm. Then we define the spectral flow along the path $\{F_t\}_{t\in[0,1]}$ to be the number:

$$\mathrm{sf}(\{F_t\}_{t\in[0,1]}) = \sum_{i=1}^{k} \mathrm{index}(P_{i-1}P_i).$$

The main results of [Phi96] show that this analytic notion is well defined being independent of the partition into 'small' intervals and that it reproduces the usual topological point of view of spectral flow. That is, we recover the intersection number approach to spectral flow when the operators in question have discrete spectrum.

The next step is to introduce spectral flow for unbounded operators. There are various ways to do this. Our definition of a Fredholm operator S in the unbounded self-adjoint case exploits the Riesz map $g : S \to S(1+S^2)^{-1/2}$, so that we say S is Fredholm if its image under this map is a bounded Fredholm operator. A path of unbounded Fredholm operators $\{S(t)\}_{t\in[0,1]}$ is said to be continuous if the path of bounded transforms $\{g(S(t))\}_{t\in[0,1]}$ is norm continuous. Then the spectral flow along $\{S(t)\}_{t\in[0,1]}$ is defined via Philips' definition to be the spectral flow along the bounded path $\{g(S(t))\}_{t\in[0,1]}$.

The analytic formulas for spectral flow that we will exploit require us to introduce differentiable paths of unbounded Fredholm operators. In the following definition we use a neutral notation in order that later the definition may be applied in different contexts.

Definition 5.1.1

(i) A path $\{D(t)\}_{t\in[0,1]}$ of unbounded self-adjoint operators is called Γ-differentiable at the point $t = t_0$ if and only if there is a bounded linear operator G such that

$$\lim_{t\to t_0} \left\| \frac{D(t)-D(t_0)}{t}(1+D(t_0)^2)^{-1/2} - G \right\| = 0.$$

In this case, we set $\dot{D}(t_0) = G(1+D(t_0)^2)^{1/2}$. By [CPS09, Lemma 25] the operator $\dot{D}(t)$ is a symmetric linear operator with the domain $\mathrm{dom}(D(t))$.

(ii) If the mapping $t \mapsto \dot{D}(t)(1+D(t)^2)^{-1/2}$, $t \in [0,1]$, is defined and continuous with respect to the operator norm, then the path $\{D(t)\}_{t\in[0,1]}$ is called continuously Γ-differentiable or a C^1_Γ-path

For our purposes we need the next result from [CPS09]:

Theorem 5.1.2 ([CPS09, Theorem 22]) *If $\{D_t\}_{t\in[0,1]}$ is a C^1_Γ-path of self-adjoint linear operators, then the path $\{g(D_t)\}_{t\in[0,1]}$ is a C^1-path with respect to the operator norm.*

We will mainly focus on spectral flow along a C^1_Γ-path $\{D_t\}_{t\in[0,1]}$ of self-adjoint unbounded Fredholm operators and using Theorem 5.1.2 we define this as

$$\mathrm{sf}(\{D_t\}_{t\in[0,1]}) := \mathrm{sf}(\{g(D_t)\}_{t\in[0,1]}). \tag{5.2}$$

5.1.1 The Variation of eta Formula

The main consequence of the original approach of [APS76] is to provide topological formulas for spectral flow. As a side issue they introduced a formula for the difference of the eta function for two Dirac type operators joined by a path along which spectral flow is defined. Melrose [Mel93] and K.P. Wojciechowski have termed this the 'variation of eta formula'. We remind the reader that the so-called 'eta invariant' is not in itself an invariant at all but forms part of the expression for the Fredholm index in the Atiyah-Patodi-Singer index theorem. There are several ways to introduce the eta function for operators with discrete spectrum but for our purposes, since we do not make explicit use of it here, we choose the simplest, namely for D self-adjoint we write

$$\eta(D) = \frac{1}{\sqrt{\pi}} \int_0^\infty \mathrm{tr}(De^{-tD^2})t^{-1/2}dt$$

whenever the trace and integral exist.

In this initial approach the definition of spectral flow involves the notion of intersection number and this is not fully explained in [APS76]. An alternative approach to the definition of spectral flow in this particular setting may be found in the book of Melrose [Mel93]. His definition, which is more or less contemporaneous with that of Phillips, uses basically the same idea.

Proposition 8.43 of [Mel93] introduces a version of the variation of eta formula, that is it gives an expression for $\eta(D_1) - \eta(D_2)$ where D_1 and D_2 are Dirac type operators corresponding to different metrics on a compact manifold that are joined by a smooth path of metrics. It requires however an explicit knowledge of the asymptotic behaviour of the one form (5.3.3) at small time. Another approach, avoiding this question of the asymptotic behaviour at time zero, appeared in the work of Getzler [Get93] in the same year as Melrose's book. Getzler uses a 'truncated eta function'

$$\eta_\epsilon(D) = \frac{1}{\sqrt{\pi}} \int_\epsilon^\infty \mathrm{tr}(De^{-tD^2})t^{-1/2}dt$$

and hence avoids the difficulty introduced by a possible singularity in the integrand at $t = 0$. Getzler's paper also explains some of the details of the intersection number definition of spectral flow that lie behind the original Atiyah-Patodi-Singer approach. A full account of Getzler's approach may also be found in the thesis of Georgescu [Geo13]. Getzler reinterprets the variation of eta formula as a formula for spectral flow and this is particularly important when studying loops of self-adjoint Fredholm operators.

We make the side remark that a vast generalisation of the original variation of eta formula may be found in the article of Dai-Zhang [DZ98] which in turn makes use of the notion of spectral section first introduced by Melrose-Piazza [MP97]. We will, however, not digress to discuss this development.

Here, in these notes, our end objective is quite different. Our aim is to relate spectral flow to scattering theory. This approach generalises beyond the Fredholm case. Nevertheless by the results of [CGK15] there is a connection to topological formulas using the small time asymptotics of the left-hand side of the principal trace formula. We will study this particular aspect by considering a specific example only (see Sect. 7.2).

5.1.2 A Review of Analytic Formulas for Spectral Flow

Next, we briefly summarise the various analytic formulas for spectral flow that have been introduced since 1993. All of these formulas for spectral flow that we discuss here have their origins in the variation of eta formula in the original Atiyah-Patodi-Singer papers [APS76]. The perspective however is different in that the variation of eta formula is now regarded as a formula for spectral flow where the difference of eta functions is just one contribution. This perspective is due to Getzler [Get93] who introduced an operator theoretic approach to earlier formulas. As we noted above a complete account of Getzler's ideas had to await the thesis of Georgescu [Geo13].

A different approach to finding an analytic formula for spectral flow may be found in [CP98, CP04] motivated by the study of spectral flow in semifinite von Neumann algebras. Though [CP98, CP04] drew inspiration from [Get93] their approach is based on Phillips' analytic viewpoint and new arguments are introduced to handle the complexities of the von Neumann algebra setting. In both of these articles formulas for spectral flow are given for paths of both bounded and unbounded Fredholm operators that may be seen to generalise the notion of a variation of eta formula.

Substantial refinements of this early work cited in the preceding paragraph are found in [ACS07] where the connection between spectral flow and the spectral shift function is undertaken in a very general framework for the first time. Though this paper and [CP98] and [CP04] involve paths of operators that may have essential spectrum there is a compactness assumption imposed that greatly restricts the applicability of the results obtained there. Some of these restrictions were lifted in [CPS09] where, within the framework of noncommutative index theory, an analytic formula for spectral flow for operators with essential spectrum (even in the standard type I case) is proved. However, the idea of studying a spectral flow formula ab initio for paths of unbounded Fredholm operators with essential spectrum and non-compact resolvent using methods from scattering theory began with [Pus08].

The reason for all of this interest in formulas for spectral flow, at least for the current authors, is their role in generalising the local index formula in non-commutative geometry due to Connes and Moscovici to the most general situation where one studies von Neumann spectral flow for paths of Breuer-Fredholm operators in semi-finite von Neumann algebras. Though we will not explore the von Neumann setting for our results in these notes we will make use of the new ideas that were introduced into non-commutative index theory in the series of papers [CGRS14, CGP+15] where the local index theorem is proved in great generality.

We remark that a refinement of [CP98, CP04] appears in [CGP+15] where yet another analytic formula for spectral flow is proved. This formula is inspired by the local index formula in non-commutative geometry and the result obtained involves taking a residue of an expression similar to the right-hand side of the principal trace formula. Taking the residue is analogous to the limiting operation used to derive an expression for spectral flow in terms of the spectral shift function that is found in this paper. We will exploit some techniques from [CGP+15] in the forthcoming discussion.

To focus now on the issues analysed here, the formula that provides the starting point for our analysis and whose history is partly outlined above comes from [CPS09]. We state the theorem for one particular instance we use in our proof.

Below the notation erf stands for the error function

$$\operatorname{erf}(x) = \frac{2}{\pi^{1/2}} \int_0^x e^{-y^2}\, dy, \quad x \in \mathbb{R}.$$

Theorem 5.1.3 ([CPS09, Theorem 9]) *Let $\{D_t\}_{t\in[0,1]}$ be a C^1_Γ-path of (unbounded) self-adjoint Fredholm operators joining endpoints D_0, D_1. Suppose that*

(i) $\int_0^1 \|\dot{D}_t e^{-\lambda D_t^2}\|_1 dt < \infty$, $\lambda > 0$;
(ii) The operator $[\frac{1}{2}\operatorname{erf}(\lambda^{1/2}D_1) - \frac{1}{2}\operatorname{erf}(\lambda^{1/2}D_0)] - [\chi_{[0,\infty)}(D_1) - \chi_{[0,\infty)}(D_0)]$ is a trace-class operator.

Then

$$\begin{aligned}\operatorname{sf}(\{D_t\}_{t\in[0,1]}) = &\int_0^1 \operatorname{tr}\big(\dot{D}_t e^{-\lambda D_t^2}\big)dt \\ &+ \operatorname{tr}\Big([\frac{1}{2}\operatorname{erf}(\lambda^{1/2}D_1) - \frac{1}{2}\operatorname{erf}(\lambda^{1/2}D_0)] - [\chi_{[0,\infty)}(D_1) - \chi_{[0,\infty)}(D_0)]\Big).\end{aligned}$$

Proof To use [CPS09, Theorem 9] set $h(x) = \sqrt{\frac{\lambda}{\pi}}e^{-\lambda x^2}$, $x \in \mathbb{R}$, $\lambda > 0$. Then clearly $\int_{\mathbb{R}} h(x)dx = 1$ and hence assumptions (i) and (ii) of [CPS09, Theorem 9] are satisfied. In addition, the antiderivative $H(\cdot)$ of $h(\cdot)$ satisfying $H(\pm\infty) = \pm\frac{1}{2}$ is given by (see [CPS09, Theorem 11])

$$H(x) = \sqrt{\frac{\lambda}{\pi}} \int_{-\infty}^x e^{-\lambda s^2} ds - \frac{1}{2} = \sqrt{\frac{\lambda}{\pi}} \int_0^x e^{-\lambda s^2} ds = \frac{1}{2}\operatorname{erf}(\sqrt{\lambda}x), \quad x \in \mathbb{R}, \lambda > 0.$$

This verifies that our assumption (ii) is precisely the assumption (iii) of [CPS09, Theorem 9]. Hence, [CPS09, Theorem 9] applies and gives the required formula. □

Remark 5.1.4 We note that [CPS09, Theorem 11] states exactly the same analytic formula for the spectral flow as we presented above. However, one of the assumptions of [CPS09, Theorem 11] requires that $e^{-\lambda D_t^2}$ is a trace-class operator for every $t \in [0, 1]$ and $\lambda > 0$. In our setting we are allowing the operators D_t to have essential spectrum so that this trace-class assumption cannot hold. ◇

The key point about the formula in the preceding theorem is that the integrand is an exact one form on the affine space of 'admissible' bounded perturbations $P(A_-)$ of A_- (what constitutes an admissible perturbation we will explain later). To understand this differential geometric viewpoint introduce the functional α on the affine space (regarded as a Banach manifold) $A_- + P(A_-)$ defined at a point $D' \in A_- + P(A_-)$ by setting

$$\alpha_{D'}(X) = \mathrm{tr}(e^{-tD'^2}X), \quad X \in P(A_-).$$

(In the context in which we use this formula we are able to prove that $e^{-tD'^2}X$ is indeed trace-class.) Then α is a one form by definition and suitably interpreting the integrand in the formula in the previous theorem we see that in the expression for spectral flow there is an integral of α along a path of perturbations of A_-.

In the historical development explained above starting with Getzler's work it is proved that this one form (in the contexts of the various papers in which it is used) is exact by various different methods. We will discuss this exactness result later using the methods of [CPS09]. In the formula for spectral flow the integral is therefore independent of the path chosen in the space of admissible perturbations. Exactness of the integrand under the assumptions we use is purely an operator theoretic fact as we show later. It does not depend on any Fredholm properties.

5.2 The Relation Between the Spectral Shift Function and the Spectral Flow

In this section we discuss the connection of the spectral shift function $\xi(\cdot; A_+, A_-)$ for the endpoints $A_\pm$ to the spectral flow along the path $\{A(t)\}_{t\in\mathbb{R}}$ in the special case when $A_\pm$ are Fredholm operators and so the spectral flow $\mathrm{sf}\{A(t)\}_{t\in\mathbb{R}}$ is well-defined. It happens that in this case, the spectral shift function at 0 computes the spectral flow modulo the correction terms. In the semi-finite setting of a semi-finite von Neumann algebra $\mathcal{M}$ equipped with faithful normal semi-finite trace τ this result was established in [ACS07] for self-adjoint operators affiliated with $\mathcal{M}$ with τ-compact resolvent. In the setting of these notes (when $\mathcal{M} = \mathcal{L}(\mathcal{H})$ and τ is the classical trace) this requires that the operators to have purely discrete spectra. Thus, the result of this section provides an extension of [ACS07] for operators with some essential spectra away from zero.

Throughout this section we assume the following:

Hypothesis 5.2.1 *In addition to Hypothesis 3.2.1 we assume that the asymptotes* $A_\pm$ *of the path* $\{A(t)\}_{t\in\mathbb{R}}$ *are Fredholm.*

Remark 5.2.2 Complex interpolation together with the Hölder inequality implies that

$$B'(t)(A_- + i)^{-1} \in \mathcal{L}_{p+1}(\mathcal{H}), \quad \int_{\mathbb{R}} \|B'(t)(A_- + i)^{-1}\|_{p+1} dt < \infty.$$

Similarly to the proof of [GLM+11, Remark 3.3] one can show that

$$B(t)(|A_-| + 1)^{-1} = \int_{-\infty}^{t} B'(s)(|A_-| + 1)^{-1}\, ds \in \mathcal{L}_{p+1}(\mathcal{H}), \quad t \in \mathbb{R}$$

In particular, $B(t), t \in \mathbb{R}$, is an A_--relatively compact operator. Hence stability of the essential spectrum implies that $A(t) = A_- + B(t), t \in \mathbb{R}$, are Fredholm operators. ◇

To prove that the spectral flow $\mathrm{sf}\{A(t)\}_{t\in\mathbb{R}}$ is well-defined we need to show that the path $\{A(t)\}_{t\in\mathbb{R}}$ is continuous in an appropriate topology. The key ingredient of our argument is Theorem 5.1.3 and that result defines the spectral flow for a C^1_Γ-path $\{S(t)\}$, $t \in [0, 1]$ (see Sect. 5.1), we re-parametrise our path $\{A(t)\}_{t=-\infty}^{\infty}$ to avoid confusion. Let $r : [0, 1] \to \mathbb{R}$ be a continuously differentiable strictly increasing function. Introduce the path $\{S(t)\}_{t=0}^{1}$ by letting

$$S(0) = A_-, \quad S(t) = A(r(t)), t \in (0, 1), \quad S(1) = A_+, \tag{5.3}$$

Lemma 5.2.3 *The path* $\{S(t)\}_{t\in[0,1]}$ *is* C^1_Γ*-path of self-adjoint Fredholm operators.*

Proof By Hypothesis 3.2.1 we have that $\{S(t)\}_{t\in[0,1]}$ is Γ-differentiable at any point and $\dot{S}(t) = A'(r(t)) \cdot r'(t) = B'(r(t)) \cdot r'(t)$. Next for arbitrary $t_1, t_2 \in [0, 1]$ we have

$$\begin{aligned}
&\left\| \dot{S}(t_1)(1 + S(t_1)^2)^{-1/2} - \dot{S}(t_2)(1 + S(t_2)^2)^{-1/2} \right\| \\
&\quad \le \|B'(r(t_1)) - B'(r(t_2))\| \|(1 + S(t_1)^2)^{-1/2}\| \\
&\quad + \|B'(r(t_2))\| \left\| (1 + S(t_1)^2)^{-1/2} - (1 + S(t_2)^2)^{-1/2} \right\|.
\end{aligned}$$

Since the family $\{B(t)\}_{t=-\infty}^{\infty}$ is continuously differentiable with respect to the uniform norm and the function r is continuous, we obtain that $\|B'(r(t_1)) -$

$B'(r(t_2))\| \to 0$ as $t_1 \to t_2$. In addition, we have

$$(1 + S(t_1)^2)^{-1/2} - (1 + S(t_2)^2)^{-1/2}$$
$$= \frac{1}{\pi}\int_0^\infty d\lambda\, \lambda^{-1/2}((1 + \lambda + S(t_1)^2)^{-1} - (1 + \lambda + S(t_2)^2)^{-1}).$$

Using the resolvent identity and continuity of the path $\{B(t)\}_{t=-\infty}^{\infty}$ one can conclude that

$$\|(1 + S(t_1)^2)^{-1/2} - (1 + S(t_2)^2)^{-1/2}\| \to 0$$

as $t_1 - t_2 \to 0$. Thus, the mapping $t \mapsto \dot{S}(t)(1 + S(t)^2)^{-1/2}$ is continuous, and hence, concludes the proof. □

Lemma 5.2.3 and (5.2) implies that the spectral flow for the path $\{A(t)\}_{t=-\infty}^{\infty}$ is well-defined by the formula

$$\mathrm{sf}(\{A(t)\}_{t=-\infty}^{\infty}) := \mathrm{sf}(\{S(t)\}_{t=0}^{1}) = \mathrm{sf}(\{g(S(t))\}_{t=0}^{1}). \tag{5.4}$$

The proof of equality (5.1) relies on the same approximation scheme as before and Theorem 5.1.3. Hence, as before (see Sect. 3.2), we introduce the family

$$A_n(t) = A_- + P_n B(t) P_n, \quad t \in \mathbb{R}, \quad A_{-,n} = A_-, \quad A_{+,n} = A_- + P_n B_+ P_n,$$

where $P_n = \chi_{[-n,n]}(A_-)$, as well as the corresponding re-parametrised path of 'cut-off' operators $\{S_n(t)\}_{t=0}^{1}$ by

$$S_n(0) = A_-, \quad S_n(t) = A_n(r(t)), t \in (0, 1), \quad S_n(1) = A_{+,n}. \tag{5.5}$$

Similarly to Lemma 5.2.3 the path $\{S_n(t)\}_{t\in\mathbb{R}}$ is a C^1_Γ-path of Fredholm operators and so the spectral flow

$$\mathrm{sf}(\{A_n(t)\}_{t=-\infty}^{\infty}) := \mathrm{sf}(\{S_n(t)\}_{t=0}^{1}) = \mathrm{sf}(\{g(S_n(t))\}_{t=0}^{1}) \tag{5.6}$$

is well-defined.

Next we check that the path $\{S_n(t)\}_{t\in[0,1]}$ satisfies the assumptions of Theorem 5.1.3. Note that $\dot{S}_n(t) = P_n B'(r(t)) P_n$ is a trace-class operator for any $t \in [0, 1]$ (see 3.6), and so $\int_0^1 \|\dot{S}_n(t)e^{-\lambda S_n(t)^2} d\|_1 t < \infty$ for all $\lambda > 0$.

Since $A_{+,n} - A_- = P_n B_+ P_n \in \mathcal{L}_1(\mathcal{H})$, Theorem 2.2.10 implies that

$$\frac{1}{2}\,\mathrm{erf}(t^{1/2}A_{+,n}) - \frac{1}{2}\,\mathrm{erf}(t^{1/2}A_-) \in \mathcal{L}_1(\mathcal{H}).$$

Furthermore, since 0 is an isolated eigenvalue of $A_{+,n}$ and A_-, there exists $\varepsilon > 0$, such that $(-\varepsilon, 0) \in \rho(A_{+,n}) \cap \rho(A_-)$. Let φ be a smooth cut-off function such that

$\varphi(t) = \chi_{(-\frac{\varepsilon}{2},\infty)}(t)$ for $t \notin (-\varepsilon, -\frac{\varepsilon}{2})$ and $\varphi' \in L_\infty(\mathbb{R})$. Using Theorem 2.2.10 again we have

$$\chi_{[0,\infty)}(A_{+,n}) - \chi_{[0,\infty)}(A_-) = \varphi(A_{+,n}) - \varphi(A_-) \in \mathcal{L}_1(\mathcal{H}).$$

Thus, the path $\{S_n(t)\}_{t\in[0,1]}$ satisfies all the assumptions of Theorem 5.1.3. Before proceeding further with our arguments we will need to recall a result from [CGP+17].

Proposition 5.2.4 *[CGP+17, Example B.6 (ii) and Theorem B.5] For the path $\{A_n(t)\}_{t\in\mathbb{R}}$ of reduced operators and $t > 0$ we have that*

$$e^{-t\boldsymbol{H}_{2,n}} - e^{-t\boldsymbol{H}_{1,n}} \in \mathcal{L}_1(L_2(\mathbb{R};\mathcal{H})), \quad \operatorname{erf}(t^{1/2}A_{+,n}) - \operatorname{erf}(t^{1/2}A_-) \in \mathcal{L}_1(\mathcal{H})$$

and the equation

$$\operatorname{tr}\left(e^{-t\boldsymbol{H}_{2,n}} - e^{-t\boldsymbol{H}_{1,n}}\right) = -\frac{1}{2}\operatorname{tr}\left(\operatorname{erf}(t^{1/2}A_{+,n}) - \operatorname{erf}(t^{1/2}A_-)\right), \tag{5.7}$$

holds.

We are now prepared for the following important result.

Proposition 5.2.5 *Let $A_{+,n}$, $\{A_n(t)\}_{t=-\infty}^{\infty}$ and $\{S_n(t)\}_{t\in[0,1]}$ be as before. Then for $\operatorname{sf}(\{A_n(t)\}_{t=-\infty}^{\infty})$, defined by (5.6) we have*

$$\begin{aligned}\operatorname{sf}(\{A_n(t)\}_{t=-\infty}^{\infty}) &= \frac{1}{2}\big[\xi(0_+; A_{+,n}, A_-) + \xi(0_-; A_{+,n}, A_-)\big] \\ &\quad + \frac{1}{2}\big[\dim(\ker(A_{+,n})) - \dim(\ker(A_-))\big].\end{aligned} \tag{5.8}$$

Proof As discussed above, the path $\{S_n(t)\}_{t\in[0,1]}$ satisfies the assumption of Theorem 5.1.3. Hence, by the result of this theorem

$$\begin{aligned}\operatorname{sf}(\{A_n(t)\}_{t=-\infty}^{\infty}) &= \operatorname{sf}(\{S_n(t)\}_{t\in[0,1]}) \\ &= \int_0^1 \operatorname{tr}\left(B_n'(t)e^{-\lambda A_n^2(r(t))}\right)dt + \frac{1}{2}\operatorname{tr}[\operatorname{erf}(\lambda^{1/2}A_{+,n}) - \operatorname{erf}(\lambda^{1/2}A_-)] \\ &\quad - \operatorname{tr}[\chi_{[0,\infty)}(A_{+,n}) - \chi_{[0,\infty)}(A_-)].\end{aligned}$$

By Proposition 2.4.6 we have

$$\int_0^1 \operatorname{tr}\left(B_n'(t)e^{-\lambda A_n^2(r(t))}\right)dt = \frac{1}{2}\operatorname{tr}[\operatorname{erf}(\lambda^{1/2}A_{+,n}) - \operatorname{erf}(\lambda^{1/2}A_-)],$$

and therefore

$$\begin{aligned}\mathrm{sf}(\{A_n(t)\}_{t=-\infty}^{\infty}) &= \mathrm{tr}[\mathrm{erf}(\lambda^{1/2}A_{+,n}) - \mathrm{erf}(\lambda^{1/2}A_-)] \\ &\quad - \mathrm{tr}[\chi_{[0,\infty)}(A_{+,n}) - \chi_{[0,\infty)}(A_-)].\end{aligned} \tag{5.9}$$

We now compute $\mathrm{tr}[\chi_{[0,\infty)}(A_{+,n}) - \chi_{[0,\infty)}(A_-)]$. Fix $\varepsilon > 0$ such that $(-\varepsilon, 0) \in \rho(A_{+,n}) \cap \rho(A_-)$. We have

$$\begin{aligned}&\mathrm{tr}\left[\chi_{[0,\infty)}(A_{+,n}) - \chi_{[0,\infty)}(A_-)\right] \\ &\quad = -\,\mathrm{tr}\left[\chi_{(-\infty,0)}(A_{+,n} - \varepsilon) - \chi_{(-\infty,0)}(A_- - \varepsilon)\right].\end{aligned}$$

Introducing the family $\tilde{B}_n(t) = B_n(t) - \varepsilon$, we have $\tilde{B}_n'(t) = B_n'(t)$ for all $t \in \mathbb{R}$, and hence the family $\{\tilde{B}_n(t)\}_{t\in\mathbb{R}}$ satisfies the conditions of [GLM+11] relative to $A_- - \varepsilon$. In addition, for the corresponding asymptotes $A_{+,n} - \varepsilon$, $A_- - \varepsilon$, 0 is not in their spectra and

$$(A_{+,n} - \varepsilon) - (A_- - \varepsilon) = A_{+,n} - A_- \in \mathcal{L}_1(\mathcal{H}).$$

Hence, by [GLM+11, Lemma 7.5] we have

$$\mathrm{tr}[\chi_{(-\infty,0)}(A_{+,n} - \varepsilon) - \chi_{(-\infty,0)}(A_- - \varepsilon)] = -\xi(0; A_{+,n} - \varepsilon, A_- - \varepsilon),$$

which implies that

$$\mathrm{tr}[\chi_{[0,\infty)}(A_{+,n}) - \chi_{[0,\infty)}(A_-)] = \xi(0; A_{+,n} - \varepsilon, A_- - \varepsilon) = \xi(\varepsilon; A_{+,n}, A_-). \tag{5.10}$$

Thus, combining (5.9) with (5.10) we conclude that

$$\mathrm{sf}(\{A_n(t)\}_{t=-\infty}^{\infty}) = \mathrm{tr}[\mathrm{erf}(\lambda^{1/2}A_{+,n}) - \mathrm{erf}(\lambda^{1/2}A_-)] - \xi(\varepsilon; A_{+,n}, A_-).$$

Next, we take the limit as $\lambda \to \infty$. Firstly, by Proposition 5.2.4 we have

$$\mathrm{sf}(\{A_n(t)\}_{t=-\infty}^{\infty}) = -\,2\,\mathrm{tr}(e^{-\lambda \boldsymbol{H}_{2,n}} - e^{-\lambda \boldsymbol{H}_{1,n}}) - \xi(\varepsilon; A_{+,n}, A_-).$$

Using [CGP+17, Theorem 4.3] it follows that

$$-2 \lim_{t\to\infty} \mathrm{tr}(e^{-\lambda \boldsymbol{H}_{2,n}} - e^{-\lambda \boldsymbol{H}_{1,n}}) = [\xi(0_+; A_{+,n}, A_-) + \xi(0_-; A_{+,n}, A_-)].$$

Therefore, since $\xi(\varepsilon; A_{+,n}, A_-) = \xi(0_+; A_{+,n}, A_-)$, we have

$$\begin{aligned}\mathrm{sf}(\{A_n(t)\}_{t=-\infty}^{\infty}) &= [\xi(0_+; A_{+,n}, A_-) + \xi(0_-; A_{+,n}, A_-)] - \xi(\varepsilon; A_{+,n}, A_-) \\ &= \frac{1}{2}[\xi(0_+; A_{+,n}, A_-) + \xi(0_-; A_{+,n}, A_-)] \\ &\quad - \frac{1}{2}[\xi(0_+; A_{+,n}, A_-) - \xi(0_-; A_{+,n}, A_-)].\end{aligned}$$

Properties of the spectral shift function (see (4.9)) imply that

$$\begin{aligned}\mathrm{sf}(\{A_n(t)\}_{t=-\infty}^{\infty}) = &\frac{1}{2}[\xi(0_+; A_{+,n}, A_-) + \xi(0_-; A_{+,n}, A_-)] \\ &+ \frac{1}{2}[\dim(\ker(A_{+,n})) - \dim(\ker(A_-))],\end{aligned}$$

which concludes the proof. □

Having established the desired formula (5.8) for the reduced operators we now want to pass to the limit as $n \to \infty$. We firstly note that Theorem 3.1.7 implies that $A_{+,n} \to A_+$ in the norm resolvent sense. Therefore, it follows from [RS80, Theorem VIII.23 (i) and (ii)] that sufficiently large $n \in \mathbb{N}$ the multiplicity of 0 for $A_{+,n}$ is the same as multiplicity for A_+. Hence,

$$\dim(\ker(A_{+,n})) = \dim(\ker(A_+)), \quad n \text{ is sufficienlty large.} \tag{5.11}$$

Recall also that the spectral shift function $\xi(\cdot; A_+, A_-)$ for the pair (A_+, A_-) is defined in (4.40) and fixed via Theorem 4.31. In addition, the operators $A_\pm$ are Fredholm, and therefore, by (4.29) we have that $\xi(\cdot; A_+, A_-)$ is left and right-continuous at zero. In addition, since $A_\pm$ and $A_{+,n}$ have discrete spectra at 0, the spectral shift functions $\xi(\cdot; A_{+,n}, A_-)$ and $\xi(\cdot; A_+, A_-)$ are step functions on a sufficiently small interval containing 0 (see (4.29)). Hence, (4.42) implies that

$$\xi(0_+; A_{+,n}, A_-) = \xi(0_+; A_+, A_-), \quad \xi(0_-; A_{+,n}, A_-) = \xi(0_-; A_+, A_-) \tag{5.12}$$

for sufficiently large $n \in \mathbb{N}$.

Next, we handle the approximation of spectral flow.

Lemma 5.2.6 *For n sufficiently large*

$$\mathrm{sf}\{A(t)\}_{t=-\infty}^{\infty} = \mathrm{sf}\{A_n(t)\}_{t=-\infty}^{\infty}.$$

Proof By (5.4) and (5.6) we have that $\mathrm{sf}\{A(t)\}_{t=-\infty}^{\infty} = \mathrm{sf}\{g(S(t))\}_{t\in[0,1]}$ and $\mathrm{sf}\{A_n(t)\}_{t=-\infty}^{\infty} = \mathrm{sf}\{g(S_n(t))\}_{t\in[0,1]}$. Recall that $S(0) = S_n(0) = A_-$, $S(1) = A_+$

and $S_n(1) = A_{+,n}$ (see (5.3) and (5.5)). We form the following loop

$$g(A_-) \longrightarrow g(A_+) \overset{\ell}{\longrightarrow} g(A_{+,n}) \longrightarrow g(A_-), \tag{5.13}$$

where the operators $g(A_+)$ and $g(A_{+,n})$ are joined by the straight line ℓ. We claim that this loop is contractible, and therefore there is no spectral flow around this loop. To this end, it is sufficient to show that all operators in the loop are compact perturbations of a fixed operator, say, $g(A_-)$.

Firstly, we show that the difference $g(A(t)) - g(A_-)$ is compact for all $-\infty \leq t \leq \infty$. By [GLM$^+$11, Lemma 6.6] we have

$$g(A(t)) - g(A_-) = T_\varphi^{A(t),A_-}\Big((A(t)^2+1)^{-1/4}\big(A(t)-A_-\big)(A_-^2+1)^{-1/4}\Big), \tag{5.14}$$

where φ is defined by setting

$$\varphi(\lambda,\mu) := \frac{\lambda(\lambda^2+1)^{-1/2} - \mu(\mu^2+1)^{-1/2}}{(\lambda^2+1)^{-1/4}\,(\lambda-\mu)\,(\mu^2+1)^{-1/4}}, \quad (\lambda,\mu)\in\mathbb{R}^2$$

and the operator $T_\varphi^{A(t),A_-}$ is bounded on $\mathcal{L}_p(\mathcal{H})$ for any $p \geq 1$. Hence, by equality (5.14) it is sufficient to show that $(A(t)^2+1)^{-1/4}\big(A(t)-A_-\big)(A_-^2+1)^{-1/4} \in \mathcal{L}_p(\mathcal{H})$ for some $p \geq 1$.

We have

$$\begin{aligned}&(A(t)^2+1)^{-1/4}\big(A(t)-A_-\big)(A_-^2+1)^{-1/4}\\ &\quad= -\overline{(A(t)^2+1)^{-1/4}(A_-^2+1)^{1/4}} \times (A_-^2+1)^{-1/4}B(t)(A_-^2+1)^{-1/4}.\end{aligned}$$

Repeating the argument in [GLM$^+$11, Remark 3.9] one can prove that the operator

$$\overline{(A(t)^2+1)^{-1/4}(A_-^2+1)^{1/4}}$$

is bounded. Using the fact that

$$B(t)(A_-^2+1)^{-1/2} \in \mathcal{L}_{p+1}(\mathcal{H}), \quad -\infty \leq t \leq \infty$$

and the three line theorem (see also [GLM$^+$11, Lemma 6.6]), we infer that $(A_-^2+1)^{-1/4}B(t)(A_-^2+1)^{-1/4} \in \mathcal{L}_{p+1}(\mathcal{H})$, which implies that

$$g(A(t)) - g(A_-) \in \mathcal{L}_{p+1}(\mathcal{H}), \quad -\infty \leq t \leq \infty.$$

Repeating the same argument, one can obtain that

$$g(A_n(t)) - g(A_-) \in \mathcal{L}_{p+1}(\mathcal{H}), \quad -\infty \leq t \leq \infty.$$

Hence, the loop (5.13) consists of compact perturbations of the operator operators $g(A_-)$, that is, it is contractible. Thus, there is no spectral flow around this loop, which means that

$$\mathrm{sf}\{g(S(t))\}_{t\in[0,1]} + \mathrm{sf}\{g(A_+), g(A_{+,n})\} + \mathrm{sf}\{g(S_n(t))\}_{t\in[1,0]} = 0. \quad (5.15)$$

Finally, by (5.11) we have $\mathrm{sf}\{g(A_+), g(A_{+,n})\} = 0$ for sufficiently large $n \in \mathbb{N}$. Hence, equality (5.15) implies that

$$\mathrm{sf}\{g(S(t))\}_{t\in[0,1]} = -\,\mathrm{sf}\{g(S_n(t))\}_{t\in[1,0]} = \mathrm{sf}\{g(S_n(t))\}_{t\in[0,1]},$$

which completes the proof. □

We now formula the main result of the present section, which relates the spectral shift function $\xi(\cdot; A_+, A_-)$ to the spectral flow $\mathrm{sf}\{A(t)\}_{t=-\infty}^{\infty}$.

Theorem 5.2.7 *Assume Hypothesis 5.2.1 then*

$$\frac{1}{2}\big(\xi(0_+; A_+, A_-) + \xi(0_-; A_+, A_-)\big)$$
$$= \mathrm{sf}\{A(t)\}_{t=-\infty}^{\infty} - \frac{1}{2}[\dim(\ker(A_+)) - \dim(\ker(A_-))].$$

Proof By Proposition 5.2.5 we have that

$$\frac{1}{2}[\xi(0_+, A_{+,n}, A_-) + \xi(0_-, A_{+,n}, A_-)]$$
$$= \mathrm{sf}(\{A_n(t)\}_{t=-\infty}^{\infty}) - \frac{1}{2}\big[\dim(\ker(A_{+,n})) - \dim(\ker(A_-))\big].$$

By (5.11) we have that $\dim(\ker(A_{+,n})) = \dim(\ker(A_+))$ for sufficiently large $n \in \mathbb{N}$. In addition, since $A_\pm$ and $A_{+,n}$ have discrete spectra at 0, the spectral shift functions $\xi(\cdot, A_{+,n}, A_-)$ and $\xi(\cdot, A_+, A_-)$ are step functions on sufficiently small interval containing 0 (see (4.10) and (4.29) respectively). Hence, Corollary 4.42 implies that $\xi(0_+, A_{+,n}, A_-) = \xi(0_+, A_+, A_-)$ and $\xi(0_-, A_{+,n}, A_-) = \xi(0_-, A_+, A_-)$ for sufficiently large $n \in \mathbb{N}$. Thus, for sufficiently large $n \in \mathbb{N}$ we have

$$\frac{1}{2}[\xi(0_+, A_+, A_-) + \xi(0_-, A_+, A_-)]$$
$$= \mathrm{sf}(\{A_n(t)\}_{t=-\infty}^{\infty}) - \frac{1}{2}\big[\dim(\ker(A_+)) - \dim(\ker(A_-))\big].$$

Referring to Lemma 5.2.6 we conclude that

$$\frac{1}{2}[\xi(0_+, A_+, A_-) + \xi(0_-, A_+, A_-)]$$
$$= \mathrm{sf}(\{A(t)\}_{t=-\infty}^{\infty}) - \frac{1}{2}\big[\dim(\ker(A_+)) - \dim(\ker(A_-))\big],$$

as required. □

5.3 Generalised Spectral Flow

We continue to use the notation and assumptions of the previous section (see Hypothesis 5.2.1). As proved in Theorem 5.2.7 the additional assumption that both asymptotes $A_\pm$ are Fredholm, guarantees that the spectral shift function $\xi(\cdot; A_+, A_-)$ at zero computes, up to correction terms the spectral flow $\mathrm{sf}\{A(t)\}_{t\in\mathbb{R}}$ along the path $\{A(t)\}_{t\in\mathbb{R}}$. When one removes the assumption that $A_\pm$ are Fredholm operators the spectral flow along a path $\{A(t)\}_{t\in\mathbb{R}}$ joining $A_\pm$, is not defined. In this section we discuss a suitable substitute for the spectral flow along the path $\{A(t)\}_{t\in\mathbb{R}}$ when the operators in the path are no longer Fredholm.

Assuming that the operators $A_\pm$ are Fredholm, consider the formula

$$\mathrm{sf}\{A(r)\}_{r=-\infty}^{\infty} = \lim_{t\uparrow\infty} -(\frac{t}{\pi})^{1/2}\int_0^1 \mathrm{tr}(e^{-tA_s^2}(A_+ - A_-))ds$$
$$+ \frac{1}{2}[\dim(\ker(A_+)) - (\dim\ker(A_-))].$$

The latter formula can be considered as a limit form of the integral formula of Theorem 5.1.3 for the spectral flow for a path of Fredholm operators $\{A(r)\}_{r=-\infty}^{\infty}$ given by a p-relative trace-class perturbation of a self-adjoint operator A_-.

This is exactly the formula, which can be considered as generalised spectral flow. That is we may now make the following:

Definition 5.3.1 Assume Hypothesis 3.2.1 and assume that the kernels of $A_\pm$ are finite-dimensional. The generalised spectral flow $\mathrm{sf}_g\{A(r)\}_{r=-\infty}^{\infty}$ of the path $\{A(r)\}_{r=-\infty}^{\infty}$ is

$$\mathrm{sf}_g\{A(r)\}_{r=-\infty}^{\infty} = \lim_{t\uparrow\infty} -(\frac{t}{\pi})^{1/2}\int_0^1 \mathrm{tr}(e^{-tA_s^2}(A_+ - A_-))ds$$
$$+ \frac{1}{2}[\dim(\ker(A_+)) - (\dim\ker(A_-))], \tag{5.16}$$

whenever the limit of the right-hand side exists.

Remark 5.3.2 Examples in low dimensions show that the limit can exist even if the operators $A_\pm$ have continuous spectrum equal to the whole real line [CGG+16, CGL+16b]. ◇

It is reasonable to ask what justifies this definition of 'generalised spectral flow'. By analogy with the properties of spectral flow in the Fredholm case [Les05] we list the corresponding properties that follow from our definition of generalised spectral flow above.

(i) We have already shown above that if the operators $A_\pm$ are Fredholm then generalised spectral flow reduces to the usual notion of spectral flow along paths of Fredholm operators.
(ii) It is obvious that this generalised spectral flow is additive on paths in the affine space of admissible perturbations by the properties of the integral on the right-hand side of (5.16).
(iii) Finally we need to have a certain path independence for this generalised spectral flow as long as paths remain in the appropriate affine space (a weak form of homotopy invariance).

This weak homotopy invariance property of generalised spectral flow needs a careful argument. It uses the fact that only p-relative trace-class perturbations of A_- are permitted. We do this in Theorem 5.3.8 by proving that the functional

$$\alpha_A(X) = \mathrm{tr}(e^{-tA^2}X), \quad A \in A_- + P(A_-), \quad X \in P(A_-),$$

where $P(A_-)$ stands for the space of p-relative trace-class perturbation of A_- gives an exact one form on the affine space $A_- + P(A_-)$.

We need to show that in the affine space of p-relative trace-class perturbations, the functional α mentioned above is well defined. As before, A_- is a self-adjoint operator and we let B be a p-relative trace-class perturbation of A_-. Since the function $u \mapsto e^{-tu^2}(u+i)^{p+1}$, $t > 0$, is bounded, the p-relative trace-class assumption on the perturbation B and Theorem 3.1.7 imply that for every $X \in P(A_-)$ the operator

$$\begin{aligned} e^{-tA_+^2}X &= (A_+ + i)^{p+1}e^{-tA_+^2}\Big((A_+ + i)^{-p-1} - (A_- + i)^{-p-1}\Big)X \\ &\quad + (A_+ + i)^{p+1}e^{-tA_+^2}\cdot(A_- + i)^{-p-1}X \end{aligned}$$

is a trace-class operator on $\mathcal{H}$. Therefore, the following definition makes sense.

Definition 5.3.3 Let A_- be a self-adjoint operator on a complex separable Hilbert space $\mathcal{H}$ and let $p \in \mathbb{N}$ be fixed. Denote by $P(A_-)$ the vector space of all bounded self-adjoint operators B on $\mathcal{H}$, which are p-relative trace-class operators with respect to A_-.

Introduce the one form α on the affine space $A_- + P(A_-)$ defined at a point $A_+ \in A_- + P(A_-)$ by setting

$$\alpha_{A_+}(X) = \mathrm{tr}(e^{-tA_+^2}X), \quad X \in P(A_-).$$

Remark 5.3.4 We note that Theorem 3.1.7 guarantees that if $B_1, B_2 \in P(A_-)$, then $B_2 \in P(A_- + B_1)$, since

$$B_2(A_- + B_1 + i)^{-p-1} = B_2(A_- + i)^{-p-1} + B_2\Big((A_- + B_1 + i)^{-p-1} - (A_- + i)^{-p-1}\Big) \in \mathcal{L}_1(\mathcal{H}).$$

◇

Theorem 5.3.8 below, shows that the form α is exact by exhibiting a function θ on the affine space $A_- + P(A_-)$ such that

$$d\theta_{A_+}(X) = \alpha_{A_+}(X), \quad A_+ \in A_- + P(A_-), \quad X \in P(A_-).$$

We follow an argument similar to the one used in [ACS07, CPS09].

Fix $B \in P(A_-)$ and denote by $A_s, s \in [0, 1]$, the straight line joining A_- and A_+, that is

$$A_s = A_- + sB, \quad s \in [0, 1].$$

By Proposition 3.1.11, we can define the function θ on the affine space $A_- + P(A_-)$ by setting

$$\theta_{A_+} = \int_0^1 \mathrm{tr}\left(Be^{-tA_s^2}\right)ds, \tag{5.17}$$

where $A_+ = A_- + B \in A_- + P(A_-)$.

Before we proceed to the proof of Theorem 5.3.8, we establish some preliminary lemmas.

Lemma 5.3.5 *Let $A_+ \in A_- + P(A_-)$ and $X \in P(A_-)$. We have*

$$\lim_{r \to 0} \int_0^1 \mathrm{tr}(Xe^{-t(A_s + srX)^2})ds = \int_0^1 \mathrm{tr}(Xe^{-tA_s^2})ds.$$

Proof The argument is similar to that in the proof of Lemma 3.1.12. By Theorem 3.1.7 (see also Remark 3.1.10) we have

$$\begin{aligned}&(A_s + srX + i)^{-j} - (A_s + i)^{-j}\\ &= \Big((A_s + srX + i)^{-j} - (A_- + i)^{-j}\Big) - \Big((A_s + i)^{-j} - (A_- + i)^{-j}\Big) \to 0\end{aligned}$$

in $\mathcal{L}_1(\mathcal{H})$ as $r \to 0$ for any fixed $j \geq p$. Therefore, Theorem 2.4.9 implies that

$$\lim_{r\to 0} \operatorname{tr}(Xe^{-t(A_s+srX)^2} - Xe^{-tA_s^2}) = 0.$$

Since, in addition, the family of functions $s \mapsto \|Xe^{-t(A_s+srX)^2}\|_1$ is uniformly bounded with respect to r by a bounded function, the dominated convergence principle implies that

$$\lim_{r\to 0} \int_0^1 \operatorname{tr}(Xe^{-t(A_s+srX)^2})ds = \int_0^1 \operatorname{tr}(Xe^{-tA_s^2})ds.$$

□

Lemma 5.3.6 *For every $j = 1, \ldots p$, $z \in \mathbb{C} \setminus \mathbb{R}$ and $X \in P(A_-)$ we have*

$$(A_{s_1} - z)^{-p+j-1} X (A_{s_2} - z)^{-j} \in \mathcal{L}_1(\mathcal{H}), \quad s_1, s_2 \in [0, 1]$$

and

$$\left\|(A_{s_1} - z)^{-p+j-1} X (A_{s_2} - z)^{-j} - (A_{s_2} - z)^{-p+j-1} X (A_{s_2} - z)^{-j}\right\|_1 \to 0$$

as $s_2 - s_1 \to 0$.

Proof We write

$$\begin{aligned}
&(A_{s_1} - z)^{-p+j-1} X (A_{s_2} - z)^{-j} \\
&\quad = \Big((A_{s_1} - z)^{-p+j-1} - (A_- - z)^{-p+j-1}\Big) X \Big((A_{s_2} - z)^{-j} - (A_- - z)^{-j}\Big) \\
&\qquad + (A_- - z)^{-p+j-1} X \cdot \Big((A_{s_2} - z)^{-j} - (A_- - z)^{-j}\Big) \\
&\qquad + \Big((A_{s_1} - z)^{-p+j-1} - (A_- - z)^{-p+j-1}\Big) \cdot X (A_- - z)^{-j} \\
&\qquad - (A_- - z)^{-p+j-1} \cdot X (A_- - z)^{-j}.
\end{aligned}$$

Hence, the claim follows from Theorem 3.1.7 and (3.3). □

Let $f(x) = e^{-tx^2}$, $x \in \mathbb{R}$, $t > 0$. Recall that the double operator integral $T_{f,a_k}^{A_s, A_s}$ is defined in Definition 2.3.6.

Lemma 5.3.7 *For the derivative* $\|\cdot\|_1\text{-}\frac{\mathrm{d}}{\mathrm{ds}}\, e^{-tA_s^2}$ *of the function* $s \mapsto e^{-tA_s^2}$, $s \in [0,1]$ *with respect to the trace norm we have*

$$\|\cdot\|_1\text{-}\frac{\mathrm{d}}{\mathrm{ds}}\, e^{-tA_s^2} = \sum_{k=1,2} T_{f,a_k}^{A_s,A_s}\Big(\sum_{j=1}^{p}(A_s - ia_k)^{-p+j-1}B(A_s - ia_k)^{-j}\Big) \in \mathcal{L}_1(\mathcal{H}).$$

Proof Let $p_0 = 2\lfloor \frac{p}{2} \rfloor + 1$. By the definition of $\|\cdot\|_1\text{-}\frac{\mathrm{d}}{\mathrm{ds}}\, e^{-tA_s^2}$ and the double operator integrals $T_{f,a_k}^{A_s,A_s}$ we have

$$\begin{aligned}
&\|\cdot\|_1\text{-}\frac{\mathrm{d}}{\mathrm{ds}}\, e^{-tA_s^2}\\
&= \|\cdot\|_1\text{-}\lim_{s_1\to s}\frac{e^{-tA_{s_1}^2} - e^{-tA_s^2}}{s_1 - s}\\
&= \|\cdot\|_1\text{-}\lim_{s_1\to s}\frac{\sum_{k=1,2} T_{f,a_k}^{A_{s_1},A_s}((A_{s_1}+ia_k)^{-p_0} - (A_s + ia_k)^{-p_0})}{s_1 - s}\\
&= \|\cdot\|_1\text{-}\lim_{s_1\to s}\frac{\sum_{k=1,2} T_{f,a_k}^{A_{s_1},A_s}(\sum_{j=1}^{p_0}(A_{s_1}+ia_k)^{-p_0+j-1}(A_s - A_{s_1})(A_s + ia_k)^{-j})}{s_1 - s}\\
&= \|\cdot\|_1\text{-}\lim_{s_1\to s}\sum_{k=1,2} T_{f,a_k}^{A_{s_1},A_s}\Big(\sum_{j=1}^{p_0}(A_{s_1}+ia_k)^{-p_0+j-1}B(A_s + ia_k)^{-j}\Big)
\end{aligned}$$

By Lemma 5.3.6 we have

$$(A_{s_1}+ia_k)^{-p_0+j-1}B(A_s + ia_k)^{-j}$$

converges to

$$(A_s+ia_k)^{-p_0+j-1}B(A_s + ia_k)^{-j}$$

in $\mathcal{L}_1(\mathcal{H})$ as $s_1 \to s$. By Theorem 3.1.7 the operators A_{s_1} and A_s satisfy Hypothesis 2.4.7. Hence, by Theorem 2.4.8 it follows that

$$T_{f,a_k}^{A_{s_1},A_s} \to T_{f,a_k}^{A_s,A_s}, \quad s_1 \to s$$

pointwise on $\mathcal{L}_1(\mathcal{H})$. Hence,

$$\|\cdot\|_1\text{-}\frac{\mathrm{d}}{\mathrm{ds}}\, e^{-tA_s^2} = \sum_{k=1,2} T_{f,a_k}^{A_s,A_s}\Big(\sum_{j=1}^{p_0}(A_s + ia_k)^{-p_0+j-1}B(A_s + ia_k)^{-j}\Big) \in \mathcal{L}_1(\mathcal{H}),$$

as required. □

The next theorem is the main result of the present section. It shows that the one form α defined in Definition 5.3.3 is exact, which implies that the right-hand side of equality (5.16) does not depend on the C^1_Γ-path joining the endpoints A_- and A_+. This provides the strongest evidence for regarding (5.16) as a generalisation of spectral flow for paths not necessarily consisting of Fredholm operators. The proof of Theorem 5.3.8 uses ideas from [ACS07, CPS09] and the approximation technique employed in the previous sections.

Theorem 5.3.8 *For every $X \in P(A_-)$ we have, for the exterior derivative:*

$$d\theta_{A_+}(X) = \alpha_{A_+}(X).$$

That is, the one form α is exact.

Proof Fix $X \in P(A_-)$ and let $B = A_+ - A_-$. By definition,

$$\begin{aligned} d\theta_{A_+}(X) &= \frac{d}{dr}\Big|_{r=0} \theta_{A_+ + rX} \\ &= \lim_{r\to 0} \frac{1}{r} \int_0^1 \operatorname{tr}\Big((A_+ + rX - A_-)e^{-t(A_s+srX)^2} - (A_+ - A_-)e^{-tA_s^2}\Big) ds \\ &= \lim_{r\to 0} \int_0^1 \operatorname{tr}(Xe^{-t(A_s+srX)^2}) ds + \lim_{r\to 0} \frac{1}{r} \int_0^1 \operatorname{tr}\big(B(e^{-t(A_s+srX)^2} - e^{-tA_s^2})\big) ds \end{aligned} \tag{5.18}$$

Now apply Lemma 5.3.5 to see that

$$\lim_{r\to 0} \int_0^1 \operatorname{tr}(Xe^{-t(A_s+srX)^2}) ds = \int_0^1 \operatorname{tr}(Xe^{-tA_s^2}) ds.$$

For the second term on the right hand side of (5.18) (in a fashion similar to [CPS09, Eq. (6)]), we claim that

$$\lim_{r\to 0} \int_0^1 \operatorname{tr}\Big(B\frac{1}{r}\big(e^{-t(A_s+srX)^2} - e^{-tA_s^2}\big)\Big) ds = \operatorname{tr}(Xe^{-tA_+^2}) - \int_0^1 \operatorname{tr}\Big(Xe^{-tA_s^2}\Big) ds. \tag{5.19}$$

Let, as before, $f(x) = e^{-tx^2}, x \in \mathbb{R}$ and $p_0 = 2\lfloor \frac{p}{2} \rfloor + 1$. With the double operator integrals defined in Definition 2.3.6 we have

$$\begin{aligned} &\frac{1}{r}B\Big(e^{-t(A_s+srX)^2} - e^{-tA_s^2}\Big) \\ &= \frac{1}{r}B \sum_{k=1,2} T_{f,a_k}^{A_s+srX,A_s}\Big((A_s + srX - ia_k)^{-p_0} - (A_s - ia_k)^{-p_0}\Big) \end{aligned}$$

$$= \frac{1}{r} B \sum_{k=1,2} T_{f,a_k}^{A_s+srX,A_s}\Big(\sum_{j=1}^{p_0}(A_s + srX - ia_k)^{-p_0+j-1} \cdot srX(A_s - ia_k)^{-j}\Big)$$

$$= sB \sum_{k=1,2} T_{f,a_k}^{A_s+srX,A_s}\Big(\sum_{j=1}^{p_0}(A_s + srX - ia_k)^{-p_0+j-1} \cdot X(A_s - ia_k)^{j}\Big).$$

Using again the convergence

$$\lim_{r\to 0}\Big\|(A_s+srX-ia_k)^{-p_0+j-1}\cdot X(A_s-ia_k)^j-(A_s-ia_k)^{-p_0+j-1}\cdot X(A_s-ia_k)^j\Big\|_1=0$$

and continuity of double operators integrals $T_{f,a_k}^{A_s+srX,A_s}$ (see Theorem 2.4.8) we obtain that

$$\lim_{r\to 0} \operatorname{tr}\Big(B\frac{1}{r}(e^{-t(A_s+srX)^2} - e^{-tA_s^2})\Big)$$
$$= s \operatorname{tr}\Big(B \sum_{k=1,2} T_{f,a_k}^{A_s,A_s}\Big(\sum_{j=1}^{p_0} (A_s - ia_k)^{-p_0+j-1} \cdot X(A_s - ia_k)^{-j}\Big)\Big).$$

We set

$$B_n = \chi_{[-n,n]}(A_s)B\chi_{[-n,n]}(A_s), \quad X_n = \chi_{[-n,n]}(A_s)X\chi_{[-n,n]}(A_s), \quad n \in \mathbb{N}.$$

As $B, X \in P(A_s)$, we have $B_n, X_n \in \mathcal{L}_1(\mathcal{H})$ for any $n \in \mathbb{N}$ (see (3.6)). Moreover, Lemma 3.1.3 combined with Lemma 5.3.6 implies that

$$\lim_{n\to\infty} \Big\|(A_s - ia_k)^{-p_0+j-1} \cdot X_n(A_s - ia_k)^{-j} - (A_s - ia_k)^{-p+j-1} \cdot X(A_s - ia_k)^{-j}\Big\|_1 = 0$$

$$\lim_{n\to\infty} \Big\|(A_s - ia_k)^{-p_0+j-1} \cdot B_n(A_s - ia_k)^{-j} - (A_s - ia_k)^{-p+j-1} \cdot B(A_s - ia_k)^{-j}\Big\|_1 = 0.$$

Hence,

$$\lim_{r\to 0} \operatorname{tr}\Big(B\frac{1}{r}(e^{-t(A_s+srX)^2} - e^{-tA_s^2})\Big)$$
$$= \lim_{n\to\infty} s \operatorname{tr}\Big(B_n \sum_{i=1,2} T_{f,a_k}^{A_s,A_s}\Big(\sum_m (A_s - ia_k)^{-p_0} \cdot X_n(A_s - ia_k)^{-m+p_0}\Big)\Big)$$
$$= \lim_{n\to\infty} s \sum_{i=1,2}\sum_m \operatorname{tr}\Big(B_n(A_s - ia_k)^{-p_0} T_{f,a_k}^{A_s,A_s}(X_n)(A_s - ia_k)^{-m+p_0}\Big)$$

$$= \lim_{n\to\infty} s \sum_{i=1,2} \sum_m \operatorname{tr}\Big((A_s - ia_k)^{-m+p_0} B_n (A_s - ia_k)^{-p_0} T^{A_s,A_s}_{f,a_k}(X_n)\Big).$$

$$= \lim_{n\to\infty} s \sum_{i=1,2} \sum_m \operatorname{tr}\Big(T^{A_s,A_s}_{f,a_k}\Big((A_s - ia_k)^{-m+p_0} B_n (A_s - ia_k)^{-p_0}\Big) X_n\Big),$$

where the last equality follows from duality and definition of double operator integrals on $\mathcal{L}(\mathcal{H})$. Therefore,

$$\lim_{r\to 0} \operatorname{tr}\Big(B\frac{1}{r}(e^{-t(A_s+srX)^2} - e^{-tA_s^2})\Big)$$
$$= s \sum_{i=1,2} \sum_m \operatorname{tr}\Big(T^{A_s,A_s}_{f,a_k}\Big((A_s - ia_k)^{-m+p_0} B (A_s - ia_k)^{-p_0}\Big) X\Big).$$

By Lemma 5.3.7 we obtain that

$$\lim_{r\to 0} \operatorname{tr}\Big(B\frac{1}{r}(e^{-t(A_s+srX)^2} - e^{-tA_s^2})\Big) = s \operatorname{tr}\Big(X\frac{d}{ds}e^{-tA_s^2}\Big).$$

Therefore,

$$\lim_{r\to 0} \int_0^1 \operatorname{tr}\Big(B\frac{1}{r}(e^{-t(A_s+srX)^2} - e^{-tA_s^2})\Big) ds = \int_0^1 s \operatorname{tr}\Big(X\frac{d}{ds}e^{-tA_s^2}\Big) ds.$$

Integrating by parts the latter equality (see also [CPS09, last display on page 1820]) we obtain that

$$\int_0^1 s \operatorname{tr}\Big(X\frac{d}{ds}e^{-tA_s^2}\Big) ds = \operatorname{tr}(Xe^{-tA_+^2}) - \int_0^1 \operatorname{tr}\Big(Xe^{-tA_s^2}\Big) ds,$$

and therefore, (5.19) is proved. Thus,

$$d\theta_{A_+}(X) = \int_0^1 \operatorname{tr}(Xe^{-tA_s^2}) ds + \operatorname{tr}(Xe^{-tA_+^2}) - \int_0^1 \operatorname{tr}\Big(Xe^{-tA_s^2}\Big) ds$$
$$= \alpha_{A_+}(X).$$

Thus, Gâteaux derivative of θ_{A_+} exists. Since it is also continuous by Theorem 2.4.9, it follows that θ_{A_+} is differentiable. Hence, $d\theta_{A+} = \alpha_{A_+}$, as required. □

Remark 5.3.9 In these notes we have not attempted to discuss what happens when we start with Phillips' spectral flow for paths in a semi-finite von Neumann algebra. It eventuates in fact that much of what we have discussed here goes through without significant change. The definition of generalised spectral flow also certainly carries over. However we do not at this time know of examples to which such a general formalism might apply. Further study of examples from condensed matter theory will almost certainly throw up situations where the semi-finite spectral flow and its generalisation appear. At the time of writing however this remains unknown territory. ◇

Chapter 6
The Principal Trace Formula and Its Applications

In this chapter we explain the proof of the fundamental result of these notes, the principal trace formula which states that (in the notation of (3.27))

$$\operatorname{tr}\left(e^{-tH_2} - e^{-tH_1}\right) = -\left(\frac{t}{\pi}\right)^{1/2} \int_0^1 \operatorname{tr}\left(e^{-tA_s^2}(A_+ - A_-)\right)ds, \quad t > 0, \tag{6.1}$$

where $A_s = A_- + s(A_+ - A_-)$, $s \in [0, 1]$ is the straight line path joining A_- and A_+ (see Theorem 6.2.2). As mentioned in the introduction, the principal trace formula allows us to establish further results for the Witten index (see Theorem 6.4.11 below) and for spectral flow (see Theorem 6.4.12).

We begin with some historical background. Then we move straight to the proof of (6.1). We also explain the proof (see Theorem 6.2.3) of the principal trace formula in its resolvent version:

$$\operatorname{tr}\left((H_2 - z)^{-k} - (H_1 - z)^{-k}\right) = -\frac{(2k-1)!!}{2^k(k-1)!} \int_0^1 \operatorname{tr}\left((A_s^2 - z)^{-\frac{1}{2}-k}(A_+ - A_-)\right)ds, \tag{6.2}$$

where, as before, $A_s = A_- + s(A_+ - A_-)$, $s \in [0, 1]$. The latter version of the principal trace formula is related to the so-called homological index (see [CK17] and [CGK16]). Our argument provides an alternative proof of [CGK16, Theorem 1.1].

Note that the occurrence of the straight line path A_s in this principal trace formula (in both versions) is an artefact of the proof. The connection to the formulation when we allow a more general path joining $A_\pm$ has already been explained in Chap. 5 where we showed that the right-hand side of the principal trace formula is independent of the path chosen to join $A_\pm$, subject of course to some simple constraints.

The remainder of this chapter is devoted to applications of the principal trace formula (6.1). Thus we first prove a generalisation of a formula due to Pushnitski that relates the spectral shift function of the pair (A_+, A_-) in (4.42) to that of the

A. Carey, G. Levitina, *Index Theory Beyond the Fredholm Case*, Lecture Notes in Mathematics 2323, https://doi.org/10.1007/978-3-031-19436-8_6

pair $(\boldsymbol{H}_2, \boldsymbol{H}_1)$. Then we move on to our main corollary which is a formula for the Witten index in terms of the spectral shift function.

The Witten index does not share the topological properties of the Fredholm index which it generalises. The issue of the sense in which the Witten index satisfies a notion of homotopy invariance was established in [CK17] and we include an overview of that result in this chapter. We conclude the discussion with a result that relates the anomaly to the spectral shift function. For the reader's convenience we summarise briefly the origin of the term anomaly in quantum physics.

6.1 A Brief History of the Principal Trace Formula

The formula that initiated the interest in finding the trace formula we prove in these notes has the form (using the notation (3.27))

$$\operatorname{tr} z\big((\boldsymbol{H}_2 - z)^{-1} - (\boldsymbol{H}_1 - z)^{-1}\big) = \frac{1}{2}\operatorname{tr}\big(g_z(A_+) - g_z(A_-)\big),\ z \in \mathbb{C}\backslash[0,\infty). \tag{6.3}$$

where $g_z(x) = x(x^2 - z)^{-1/2}$.

This formula was initially proved in [BGG+87] in the special case when the operators A_+ and A_- act on a one-dimensional Hilbert space. To the best of our knowledge, the formula (6.3) was proved in the infinite-dimensional case, without any assumption of discreteness of the spectra of $A_\pm$, by Pushnitski in [Pus08] in the setting where the path $\{B'(t)\}_{t\in\mathbb{R}}$ consists of trace-class operators. An extension of (6.3) appeared in the paper [GLM+11] under a relative trace-class perturbation assumption. The assumptions of [Pus08] and [GLM+11] guaranteed that operators on both sides of (6.3) are trace-class operators. In retrospect, due to Proposition 2.4.6, the trace formula (6.3) is a special case of the principal trace formula (6.2).

The clue to the generalisation of the trace formula of [GLM+11] lies in a trace formula that was proved in [BCP+06] in the setting of the Atiyah L^2-index theorem for Dirac-type operators on covering spaces of compact manifolds. The set-up there is that we have a simply connected manifold $\tilde{M}$ admitting an action by a discrete group Γ such that the quotient $M = \tilde{M}/\Gamma$ is a compact manifold. The operators $A_\pm$ considered in this setting are of Dirac type acting on sections of a vector bundle over $\tilde{M}$. They are assumed to be unitarily equivalent (by a bundle automorphism) and Γ-invariant. We introduce the notation τ_Γ for the Γ trace of Atiyah remarking that this trace gives rise to a semifinite von Neumann algebra that is not type I in general.

In [BCP+06] the interest was in the 'von Neumann spectral flow' along a path $A(s), s \in [0,1]$ of Γ invariant self-adjoint Dirac type operators joining $A_\pm$. This is defined by taking Phillips analytic approach to spectral flow (see Chap. 5) and generalising his definition so that it applies in a semifinite von Neumann algebra (this is established in [BCP+06]). Note that in this semifinite von Neumann algebra setting operators may have continuous spectrum including zero but still be 'Breuer-Fredholm' as explained in [BCP+06].

We may form $\boldsymbol{D}_A$ as before and regard this as a differential operator on a Hilbert space $\mathcal{K}$ of sections of a bundle on $S^1 \times \tilde{M}$. There is also a Γ trace defined on operators on $\mathcal{K}$ denoted $\tau_{\Gamma,\mathcal{K}}$.

Then the following trace formula is shown to hold in [BCP+06]:

$$\tau_{\Gamma,\mathcal{K}}(e^{-\epsilon \boldsymbol{D}_A^* \boldsymbol{D}_A} - e^{-\epsilon \boldsymbol{D}_A \boldsymbol{D}_A}) = \sqrt{\frac{\epsilon}{\pi}} \int \tau_\Gamma(\frac{dA(s)}{ds} e^{-\epsilon A(s)^2}) ds$$

The left-hand side is the (Breuer)-Fredholm index and the right-hand side is semi-finite spectral flow.

There is a resolvent version of this formula that we now explain. Using a Laplace transform argument we obtain for $r > 0$: on the left-hand side:

$$\int_0^\infty \epsilon^r e^{-\epsilon}(e^{-\epsilon \boldsymbol{D}_A^* \boldsymbol{D}_A} - e^{-\epsilon \boldsymbol{D}_A \boldsymbol{D}_A^*}) d\epsilon = \Gamma(r)((1 + \boldsymbol{D}_A^* \boldsymbol{D}_A)^{-r} - (1 + \boldsymbol{D}_A \boldsymbol{D}_A^*)^{-r}),$$

while, inserting the same function of ϵ and integrating on the right-hand side we obtain:

$$\int_0^\infty \epsilon^{r+1/2} e^{-\epsilon} \pi^{-1/2} \int \frac{dA(s)}{ds} e^{-\epsilon A(s)^2} ds d\epsilon.$$

The order of integration may be interchanged, and so we obtain

$$\Gamma(r + 1/2)\pi^{-1/2} \int \frac{dA(s)}{ds}(1 + A(s)^2)^{-(r+1/2)} ds.$$

So we have on each side of the formula respectively:

$$\Gamma(r)((1 + \boldsymbol{D}_A^* \boldsymbol{D}_A)^{-r} - (1 + \boldsymbol{D}_A \boldsymbol{D}_A^*)^{-r})$$

and

$$\Gamma(r + 1/2)\pi^{-1/2} \int \frac{dA(s)}{ds}(1 + A(s)^2)^{-(r+1/2)} ds.$$

Under the assumption that all the resolvents in both expressions are trace-class and under appropriate continuity assumptions for the path $A(s)$ we may take traces and then interchange the trace and integral to obtain the following trace formula.

$$\tau_{\Gamma,\mathcal{K}}((1 + \boldsymbol{D}_A^* \boldsymbol{D}_A)^{-r} - (1 + \boldsymbol{D}_A \boldsymbol{D}_A^*)^{-r}) = C_{r+1/2} \int \tau_\Gamma(\frac{dA(s)}{ds}(1 + A(s)^2)^{-(r+1/2)}) ds$$

where $C_{r+1/2} = \Gamma(r + 1/2)\pi^{-1/2}\Gamma(r)^{-1}$.

Chronologically it is this resolvent formula that was first proved to hold without compact resolvent assumptions. In fact, in [CGK16] a purely operator theoretic version of this trace formula that makes no Fredholm assumptions on A_- was proved. The approach of that paper was not inspired by non-commutative geometry but by the pseudo-differential operator methods in [CGK15]. However the formula itself motivated us to prove the fundamental result of these notes.

6.2 Proving the Principal Trace Formula

In this section we explain the proof Theorem 6.2.2. This result provides us with the principal trace formula expressing the trace of the semigroup difference $e^{-t\boldsymbol{H}_2} - e^{-t\boldsymbol{H}_1}$ on the Hilbert space $L_2(\mathbb{R};\mathcal{H})$ via the trace of an operator on the 'inner' Hilbert space $\mathcal{H}$.

Our approach to the proof of the principal trace formula relies on an approximation approach in which we use results already known for the path $\{A_n(t)\}_{t\in\mathbb{R}}$ of reduced operators

$$A_n(t) = A_- + P_n B(t) P_n,$$

where, as before, $P_n = \chi_{[-n,n]}(A_-)$.

Now recall the statement of Proposition 5.2.4. Namely, for the path $\{A_n(t)\}_{t\in\mathbb{R}}$ of reduced operators we have that

$$e^{-t\boldsymbol{H}_{2,n}} - e^{-t\boldsymbol{H}_{1,n}} \in \mathcal{L}_1(L_2(\mathbb{R};\mathcal{H})), \quad \operatorname{erf}(t^{1/2}A_{+,n}) - \operatorname{erf}(t^{1/2}A_-) \in \mathcal{L}_1(\mathcal{H})$$

and the equation

$$\operatorname{tr}\left(e^{-t\boldsymbol{H}_{2,n}} - e^{-t\boldsymbol{H}_{1,n}}\right) = -\frac{1}{2}\operatorname{tr}\left(\operatorname{erf}(t^{1/2}A_{+,n}) - \operatorname{erf}(t^{1/2}A_-)\right), \tag{6.4}$$

holds.

Remark 6.2.1 One may think that we can now approximate both sides of Eq. (6.4) and obtain a more general version of the formula obtained in [CGP+17] (see also [BCP+06]). However, as shown in Example 4.2.8 this is not true in the generality we assume here. Indeed, for the Dirac operator $A_- = \mathcal{D}$ and $A_+ = \mathcal{D} + V$ on $\mathbb{R}^d$ with sufficiently good potential V, the operator $\operatorname{erf}(t^{1/2}A_+) - \operatorname{erf}(t^{1/2}A_-)$ is not a trace-class operator. Thus, on the right-hand side of (6.4) we cannot pass to the limit, in general. ◇

To overcome the obstacle mentioned in the above remark we firstly use the (noncommutative) Fundamental Theorem of Calculus obtained in Proposition 2.4.6 applied to the right-hand side of (6.4) and only after that pass to the limit as $n \to \infty$.

Since the operator $B_{+,n} = A_{+,n} - A_-$ is a trace-class operator (see (3.6)), it follows that the path $B_{s,n} = s(A_{+,n} - A_-)$ is a C^1-path of trace-class operators. Applying now Proposition 2.4.6 for this path (with $f = \mathrm{erf}$, which clearly satisfies the assumption of this proposition since f' is a Schwartz function) we obtain that

$$\frac{1}{2}\operatorname{tr}\Big(\operatorname{erf}(t^{1/2}A_{+,n}) - \operatorname{erf}(t^{1/2}A_-)\Big) = \Big(\frac{t}{\pi}\Big)^{1/2}\int_0^1 \operatorname{tr}\Big(e^{-tA_{s,n}^2}(A_{+,n} - A_-)\Big)ds.$$

Hence, Eq. (6.4) can be rewritten as

$$\operatorname{tr}\Big(e^{-t\boldsymbol{H}_{2,n}} - e^{-t\boldsymbol{H}_{1,n}}\Big) = -\Big(\frac{t}{\pi}\Big)^{1/2}\int_0^1 \operatorname{tr}\Big(e^{-tA_{s,n}^2}(A_{+,n} - A_-)\Big)ds, \tag{6.5}$$

We are now ready to prove the principal trace formula in its heat kernel version, which is the main result of this chapter.

Theorem 6.2.2 (The Principal Trace Formula) *Assume Hypothesis 3.2.5. Let* $A_s = A_- + s(A_+ - A_-)$, $s \in [0, 1]$, *be the straight line path joining* A_- *and* A_+. *Then for all* $t > 0$, *we have*

$$\operatorname{tr}\Big(e^{-t\boldsymbol{H}_2} - e^{-t\boldsymbol{H}_1}\Big) = -\Big(\frac{t}{\pi}\Big)^{1/2}\int_0^1 \operatorname{tr}\Big(e^{-tA_s^2}(A_+ - A_-)\Big)ds.$$

Proof By (6.5) we have

$$\operatorname{tr}\Big(e^{-t\boldsymbol{H}_{2,n}} - e^{-t\boldsymbol{H}_{1,n}}\Big) = -\Big(\frac{t}{\pi}\Big)^{1/2}\int_0^1 \operatorname{tr}\Big(e^{-tA_{s,n}^2}(A_{+,n} - A_-)\Big)ds. \tag{6.6}$$

We now pass to the limit as $n \to \infty$.

For the left hand side of (6.6) we firstly note that Theorem 3.2.6 guarantees that the assumptions of Theorem 2.4.9 are satisfied with $A_n = \boldsymbol{H}_{2,n}$, $A = \boldsymbol{H}_2$, $B_n = \boldsymbol{H}_{1,n}$, $B = \boldsymbol{H}_1$. Hence, since the function $f(\lambda) = e^{-t\lambda^2}$, $t > 0$, is a Schwartz function (and so, in particular, by (2.19) it belongs to the class $\mathfrak{F}_m(\mathbb{R})$ for any $m \in \mathbb{N}$), it follows from Theorem 2.4.9 that

$$\lim_{n\to\infty} \operatorname{tr}\Big(e^{-t\boldsymbol{H}_{2,n}} - e^{-t\boldsymbol{H}_{1,n}}\Big) = \operatorname{tr}\Big(e^{-t\boldsymbol{H}_2} - e^{-t\boldsymbol{H}_1}\Big). \tag{6.7}$$

For the right hand side of (6.6) by Proposition 3.1.12 we have

$$\lim_{n\to\infty}\int_0^1 \operatorname{tr}\Big(e^{-tA_{s,n}^2}(A_{+,n} - A_-)\Big)ds = \int_0^1 \operatorname{tr}\Big(e^{-tA_s^2}(A_+ - A_-)\Big)ds.$$

Thus, (6.6) and (6.7) imply that

$$\operatorname{tr}\left(e^{-t\boldsymbol{H}_2} - e^{-t\boldsymbol{H}_1}\right) = -\left(\frac{t}{\pi}\right)^{1/2} \int_0^1 \operatorname{tr}\left(e^{-tA_s^2}(A_+ - A_-)\right) ds,$$

which concludes the proof. □

To conclude this section we show how our technique gives an alternative proof of the resolvent version of the principal trace formula proved in [CGK16]. The resolvent version of the formula is known to be connected to cyclic homology (see references in [CGK16]). It is likely that further homological properties of the principal trace formula may be established using the results of that paper.

Theorem 6.2.3 *Assume Hypothesis 3.2.5. Let $A_s = A_- + s(A_+ - A_-)$, $s \in [0, 1]$ be the straight line path joining A_- and A_+. Then for all $k \geq m$ and all $z < 0$, we have that*

$$\operatorname{tr}\left((\boldsymbol{H}_2 - z)^{-k} - (\boldsymbol{H}_1 - z)^{-k}\right) = -\frac{(2k-1)!!}{2^k(k-1)!} \int_0^1 \operatorname{tr}\left((A_s^2 - z)^{-\frac{1}{2}-k}(A_+ - A_-)\right) ds.$$

Proof Fix $z < 0$ and consider the function

$$f(\lambda) = (\lambda - z)^{-k}, \quad \lambda \in [0, \infty), k \in \mathbb{N}.$$

Then for the function F on $\mathbb{R}$, defined by (cf. [CGP⁺17, (B.57)])

$$F(\nu) = \frac{\nu}{2\pi} \int_{[\nu^2, \infty)} \lambda^{-1}(\lambda - \nu^2)^{-1/2}[f(\lambda) - f(0)] d\lambda,$$

we have (see [CGP⁺17, (B.51)])

$$F'(\nu) = \frac{1}{\pi} \int_{[0,\infty)} \lambda^{-1/2} f'(\lambda + \nu^2) d\lambda, \quad \nu \in \mathbb{R}.$$

For this choice of f, one may compute that

$$\begin{aligned}
F'(\nu) &= -\frac{k}{\pi} \int_{[0,\infty)} \frac{1}{\lambda^{1/2}(\lambda - z + \nu^2)^{k+1}} d\lambda \\
&= \frac{k}{\pi} \frac{\Gamma(k + 1/2)\sqrt{\pi}}{\Gamma(k+1)} \frac{1}{(\nu^2 - z)^{k+1/2}} \\
&= -\frac{(2k-1)!!}{2^k(k-1)!} \frac{1}{(\nu^2 - z)^{k+1/2}},
\end{aligned}$$

where the second equality follows from the definition of the Beta function and the last equality from properties of the Gamma function.

Hence, by Carey et al. [CGP+17, Theorem B.5] and Proposition 2.4.6 we have

$$\begin{aligned}
&\operatorname{tr}\Big((\boldsymbol{H}_{2,n}-z)^{-k}-(\boldsymbol{H}_{1,n}-z)^{-k}\Big)\\
&\quad=\operatorname{tr}\big(F(A_{+,n})-F(A_-)\big)\\
&\quad=\int_0^1 \operatorname{tr}\big(F'(A_{s,n})(A_{+,n}-A_-)\big)ds\\
&\quad=-\frac{(2k-1)!!}{2^k(k-1)!}\int_0^1 \operatorname{tr}\big((A_{s,n}^2-z)^{-k-1/2}(A_{+,n}-A_-)\big)ds.
\end{aligned}$$

Passing to the limit as $n\to\infty$ as before, we infer the principal trace formula in the resolvent form. □

Remark 6.2.4 Our technique may be used to prove trace formulas for a wide class of functions thus going beyond resolvents and heat kernels. However, at this point it is not clear that there is a use for formulas for a general class of functions, and so we omit this refinement. ◇

6.3 A Generalised Pushnitski Formula

Theorem 6.3.1 *Assume Hypothesis 3.2.5. Let $\xi(\cdot;A_+,A_-)$ be the spectral shift function for the pair (A_+,A_-) fixed in* (4.42) *and let $\xi(\cdot;\boldsymbol{H}_2,\boldsymbol{H}_1)$ be the spectral shift function for the pair $(\boldsymbol{H}_2,\boldsymbol{H}_1)$ fixed by equality* (4.38). *Then for a.e. $\lambda>0$ we have*

$$\xi(\lambda;\boldsymbol{H}_2,\boldsymbol{H}_1)=\frac{1}{\pi}\int_{-\lambda^{1/2}}^{\lambda^{1/2}}\frac{\xi(\nu;A_+,A_-)\,d\nu}{(\lambda-\nu^2)^{1/2}} \tag{6.8}$$

with a convergent Lebesgue integral on the right-hand side of (6.8).

Proof By the principle trace formula for the semigroup difference (see Theorem 6.2.2) we have that

$$\operatorname{tr}(e^{-t\boldsymbol{H}_2}-e^{-t\boldsymbol{H}_1})=-\Big(\frac{t}{\pi}\Big)^{1/2}\int_0^1 \operatorname{tr}\big(e^{-tA_s^2}(A_+-A_-)\big)ds$$

for all $t > 0$. On the right hand side of this formula using (3.14) and (6.2) we have that

$$\begin{aligned}
&\left(\frac{t}{\pi}\right)^{1/2} \int_0^1 \operatorname{tr}(e^{-tA_s^2}(A_+ - A_-))ds \\
&\overset{(3.14)}{=} \lim_{n\to\infty} \left(\frac{t}{\pi}\right)^{1/2} \int_0^1 \operatorname{tr}(e^{-tA_{s,n}^2}(A_{+,n} - A_-))ds \qquad (6.9)\\
&\overset{(6.2)}{=} \lim_{n\to\infty} \frac{1}{2} \operatorname{tr}(\operatorname{erf}(t^{1/2}A_{+,n}) - \operatorname{erf}(t^{1/2}A_-)).
\end{aligned}$$

By Lipschitz-Krein trace formula (see Theorem 4.1.3) and the definition of the error function it follows that

$$\frac{1}{2}\operatorname{tr}\left(\operatorname{erf}(t^{1/2}A_{+,n}) - \operatorname{erf}(t^{1/2}A_-)\right) = \left(\frac{t}{\pi}\right)^{1/2} \int_{\mathbb{R}} e^{-ts^2}\xi(s; A_{+,n}, A_-)ds. \qquad (6.10)$$

By Corollary 4.3.6 we obtain

$$\lim_{n\to\infty} \left(\frac{t}{\pi}\right)^{1/2} \int_{\mathbb{R}} e^{-ts^2}\xi(s; A_{+,n}, A_-)ds = \left(\frac{t}{\pi}\right)^{1/2} \int_{\mathbb{R}} e^{-ts^2}\xi(s; A_+, A_-)ds. \qquad (6.11)$$

Thus, combining (6.9), (6.10) and (6.11) we conclude that the right-hand side of the principal trace formula can be written as

$$\left(\frac{t}{\pi}\right)^{1/2} \int_0^1 \operatorname{tr}(e^{-tA_s^2}(A_+ - A_-))ds = \left(\frac{t}{\pi}\right)^{1/2} \int_{\mathbb{R}} e^{-ts^2}\xi(s; A_+, A_-)ds.$$

We use the function $s \mapsto e^{-ts}$, $s \in (a, \infty)$, $t > 0$ and extend it to all of $\mathbb{R}$ by patching on a function of compact support decreasing smoothly to zero on $(-\infty, a]$ so that it belongs to the class $\mathfrak{F}_m(\mathbb{R})$ (see (2.19)). It will be clear below how to choose a appropriately. Hence, by Lipschitz-Krein trace formula (4.39) for the left hand side of the principal trace formula we have that

$$\operatorname{tr}(e^{-t\boldsymbol{H}_2} - e^{-t\boldsymbol{H}_1}) = -t \int_0^\infty \xi(\lambda; \boldsymbol{H}_2, \boldsymbol{H}_1)e^{-t\lambda}\, d\lambda.$$

Thus,

$$\begin{aligned}
\int_0^\infty \xi(\lambda; \boldsymbol{H}_2, \boldsymbol{H}_1)e^{-t\lambda}\, d\lambda &= \left(\frac{1}{\pi\cdot t}\right)^{1/2} \int_{\mathbb{R}} \xi(s; A_+, A_-)e^{-ts^2}ds \\
&= \left(\frac{1}{\pi\cdot t}\right)^{1/2} \int_0^\infty \frac{\xi(\sqrt{s}; A_+, A_-) + \xi(-\sqrt{s}; A_+, A_-)}{\sqrt{s}} e^{-ts}ds, \qquad (6.12)
\end{aligned}$$

where for the last integral we used the substitutions $s \mapsto \sqrt{s}$ and $s \mapsto -\sqrt{s}$ for the integrals on $(0, \infty)$ and on $(-\infty, 0)$, respectively.

Let us denote by L the Laplace transform on $L^1_{loc}(\mathbb{R})$. It is well-known that $L(\frac{1}{\pi\sqrt{s}})(t) = \frac{1}{\sqrt{\pi t}}$ (see e.g. [AS64, 29.3.4]). Therefore, introducing

$$\xi_0(s) := \frac{\xi(\sqrt{s}; A_+, A_-) + \xi(-\sqrt{s}; A_+, A_-)}{\sqrt{s}}, \quad s \in [0, \infty),$$

equality (6.12) can be rewritten as

$$L\big(\xi(\lambda; \boldsymbol{H}_2, \boldsymbol{H}_1)\big)(t) = L\Big(\frac{1}{\pi\sqrt{s}}\Big)(t) \cdot L\big(\xi_0(s)\big)(t).$$

By Arendt et al. [ABHN01, Proposition 1.6.4] the right-hand side of the above equality is equal to $L\Big(\frac{1}{\pi\sqrt{s}} * \xi_0(s)\Big)(t)$. Therefore, by the uniqueness theorem for the Laplace transform (see e.g. [ABHN01, Theorem 1.7.3]) we have $\xi(\lambda; \boldsymbol{H}_2, \boldsymbol{H}_1) = \big(\frac{1}{\pi\sqrt{s}} * \xi_0(s)\big)(\lambda)$ for a.e. $\lambda \in [0, \infty)$. Thus, for a.e. $\lambda \in [0, \infty)$ we have

$$\begin{aligned}
\xi(\lambda; \boldsymbol{H}_2, \boldsymbol{H}_1) &= \frac{1}{\pi} \int_0^\lambda \frac{1}{\sqrt{\lambda - s}} \xi_0(s) ds \\
&= \frac{1}{\pi} \int_0^\lambda \frac{1}{\sqrt{\lambda - s}} \frac{\xi(\sqrt{s}; A_+, A_-) + \xi(-\sqrt{s}; A_+, A_-)}{\sqrt{s}} ds \\
&= \frac{1}{\pi} \int_0^\lambda \frac{\xi(\sqrt{s}; A_+, A_-)}{\sqrt{s}\sqrt{\lambda - s}} ds + \frac{1}{\pi} \int_0^\lambda \frac{\xi(-\sqrt{s}; A_+, A_-)}{\sqrt{s}\sqrt{\lambda - s}} ds \\
&= \frac{1}{\pi} \int_{-\sqrt{\lambda}}^{\sqrt{\lambda}} \frac{\xi(s; A_+, A_-) ds}{\sqrt{\lambda - s^2}},
\end{aligned}$$

as required. □

6.4 The Witten Index

6.4.1 Preliminaries

In his paper [Wit82], Witten introduced a number, which counts the difference in the number of bosonic and fermionic zero-energy modes of a Hamiltonian. This quantity, called the Witten index, became popular in connection with a variety of examples in supersymmetric quantum mechanics in the 1980s and in [BGG+87, BGGS87, GS88] has been put in mathematical framework using two different regularisation, which we recall next.

We start with the following facts on trace-class properties of resolvent and semigroup differences.

Proposition 6.4.1 *Suppose that $0 \leq S_j$, $j = 1, 2$, are nonnegative, self-adjoint operators in $\mathcal{H}$.*

(i) (see e.g. [Wei80, p. 178]) If $\left[(S_2 - z_0)^{-1} - (S_1 - z_0)^{-1}\right] \in \mathcal{L}_1(\mathcal{H})$ for some $z_0 \in \rho(S_1) \cap \rho(S_2)$, then

$$\left[(S_2 - z)^{-1} - (S_1 - z)^{-1}\right] \in \mathcal{L}_1(\mathcal{H}) \; for \; all \; z \in \rho(S_1) \cap \rho(S_2).$$

(ii) (see e.g. [CGP+17, Lemma 3.1]) If $\left[e^{-t_0 S_2} - e^{-t_0 S_1}\right] \in \mathcal{L}_1(\mathcal{H})$ for some $t_0 > 0$, then

$$\left[e^{-tS_2} - e^{-tS_1}\right] \in \mathcal{L}_1(\mathcal{H}) \; for \; all \; t \geq t_0.$$

The preceding fact allows one to consider the following two definitions.

Let T be a closed, linear, densely defined operator in $\mathcal{H}$. Suppose that for some $t_0 > 0$

$$\left[e^{-t_0 T^*T} - e^{-t_0 TT^*}\right] \in \mathcal{L}_1(\mathcal{H}).$$

Then $\left(e^{-tT^*T} - e^{-tTT^*}\right) \in \mathcal{L}_1(H)$ for all $t > t_0$ and one introduces the semigroup regularization

$$\Delta_s(T, t) = \operatorname{tr}\left(e^{-tT^*T} - e^{-tTT^*}\right), \quad t > 0. \tag{6.13}$$

Definition 6.4.2 *The semigroup regularized Witten index $W(T)$ of T is defined by*

$$W_s(T) = \lim_{t \uparrow \infty} \Delta_s(T, t),$$

whenever this limit exists.

Similarly, suppose that for some (and hence for all) $z \in \mathbb{C} \backslash [0, \infty) \subseteq [\rho(T^*T) \cap \rho(TT^*)]$,

$$\left[(T^*T - z)^{-1} - (TT^* - z)^{-1}\right] \in \mathcal{L}_1(\mathcal{H}).$$

Then one introduces the resolvent regularization

$$\Delta_r(T, \lambda) = (-\lambda) \operatorname{tr}\left((T^*T - \lambda)^{-1} - (TT^* - \lambda)^{-1}\right), \quad \lambda < 0. \tag{6.14}$$

Definition 6.4.3 *The resolvent regularized Witten index $W_r(T)$ of T is defined by*

$$W_r(T) = \lim_{\lambda \uparrow 0} \Delta_r(T, \lambda),$$

whenever this limit exists.

As proved in [GS88, BGG+87], the Witten index of an operator T in $\mathcal{H}$ is a natural substitute for the Fredholm index of T in cases where the operator T ceases to have the Fredholm property. Namely, the following result states that both (resolvent and semigroup) regularized Witten indices coincide with the Fredholm index in the special case of Fredholm operators.

Theorem 6.4.4 ([GS88, BGG+87]) *Let T be an (unbounded) Fredholm operator in H. Suppose that $\left[(T^*T - z)^{-1} - (TT^* - z)^{-1}\right]$, $\left[e^{-t_0 T^*T} - e^{-t_0 TT^*}\right] \in \mathcal{L}_1(\mathcal{H})$ for some $z \in \mathbb{C}\backslash[0, \infty)$, and $t_0 > 0$. Then*

$$\text{index}(T) = W_r(T) = W_s(T).$$

We note that the regularisations (6.14) and (6.13) has been used before [GS88, BGG+87] to compute the Fredholm index of an operator (see e.g. [Cal78]).

In general (i.e., if T is not Fredholm), $W_r(T)$ (respectively, $W_s(T)$) is not necessarily integer-valued; in fact, it can be any real number. As a concrete example, we mention the two-dimensional magnetic field example discussed by Aharonov and Casher [AC79] which demonstrates that the resolvent and semigroup regularized Witten indices are equal to the (non-quantized) magnetic flux $F \in \mathbb{R}$ which indeed can be any prescribed real number.

Expressing the Witten index $W_s(T)$ (respectively, $W_r(T)$) of an operator T in terms of the spectral shift function $\xi(\,\cdot\,; T^*T, TT^*)$ requires of course the choice of a concrete representative of the spectral shift function:

Theorem 6.4.5 ([BGG+87, GS88])

(i) Suppose that $\left[e^{-t_0 T^*T} - e^{-t_0 TT^*}\right] \in \mathcal{L}_1(\mathcal{H})$ for some $t_0 > 0$ and the spectral shift function $\xi(\,\cdot\,; T^*T, TT^*)$, uniquely defined by the requirement $\xi(\lambda; T^*T, TT^*) = 0$, $\lambda < 0$, is continuous from above at $\lambda = 0$. Then the semigroup regularized Witten index $W_s(T)$ of T exists and

$$W_s(T) = -\xi(0_+; T^*T, TT^*).$$

(ii) Suppose that $\left[(T^*T - z)^{-1} - (TT^* - z)^{-1}\right] \in \mathcal{L}_1(\mathcal{H})$, $z \in \mathbb{C}\backslash[0, \infty)$ and $\xi(\,\cdot\,; T^*T, TT^*)$, uniquely defined by the requirement $\xi(\lambda; T^*T, TT^*) = 0$, $\lambda < 0$, is bounded and piecewise continuous on $\mathbb{R}$. Then the resolvent regularized Witten index $W_r(T)$ of T exists and

$$W_r(T) = -\xi(0_+; T^*T, TT^*).$$

In our setting we aim to consider the case when T is a differential type operator. In this case, it is typical that the difference of resolvents $(T^*T-z)^{-1}-(TT^*-z)^{-1}$ belongs to a higher Schatten class as would be expected for the study of differential operators in higher dimensions. Therefore, we need to modify the definition of resolvent regularisation of the Witten index.

Definition 6.4.6 Let T be a closed, linear, densely defined operator acting in $\mathcal{H}$ and let $k \in \mathbb{N}$. Suppose that for all $\lambda < 0$ we have that

$$\left[(T^*T-\lambda)^{-k}-(TT^*-\lambda)^{-k}\right] \in \mathcal{L}_1(\mathcal{H}).$$

Then the k-th resolvent regularized Witten index $W_{k,r}(T)$ of T is defined by

$$W_{k,r}(T) = \lim_{\lambda \uparrow 0} (-\lambda)^k \operatorname{tr}\left((T^*T-\lambda)^{-k}-(TT^*-\lambda)^{-k}\right) \tag{6.15}$$

whenever this limit exists.

Remark 6.4.7 We recall that the k-th resolvent regularised Witten index is the limit (as $\lambda \uparrow 0$) of the so-called homological index (see [CGK15, CGK16]). ◇

6.4.2 *The Formula in Terms of the Spectral Shift Function*

In this subsection we prove Theorem 1.3.1, which provides a relation between the Witten index of the operator $\boldsymbol{D}_{\boldsymbol{A}}$ and the spectral shift function $\xi(\cdot; A_+, A_-)$. We follow the detailed treatment in [CGP+17]. Related earlier work used a 'relatively trace-class perturbation assumption' in [CGP+17] and then a 'relatively Hilbert-Schmidt class perturbation assumption' in [CGLS16a]. The formula described here implies that the old difficulty, that the only examples for which the Witten index was defined were low dimensional, is removed. In Sect. 7.1 we show that our Hypothesis 3.2.1 permits consideration of differential operators (in particular, Dirac operators) in any dimension uniformly. To relate the Witten index of the operator $\boldsymbol{D}_{\boldsymbol{A}}$ to the spectral shift function for the pair (A_+, A_-), we first recall some necessary definitions.

Definition 6.4.8 Let $f \in L_{1,loc}(\mathbb{R})$ and $h > 0$.

(i) The point $x \in \mathbb{R}$ is called a right Lebesgue point of f if there exists an $\alpha_+ \in \mathbb{C}$ such that

$$\lim_{h \downarrow 0} \frac{1}{h} \int_x^{x+h} |f(y) - \alpha_+| dy = 0.$$

One then denotes $\alpha_+ = f_L(x_+)$.

(ii) The point $x \in \mathbb{R}$ is called a left Lebesgue point of f if there exists an $\alpha_- \in \mathbb{C}$ such that

$$\lim_{h\downarrow 0} \frac{1}{h} \int_{x-h}^{x} |f(y) - \alpha_-|dy = 0.$$

One then denotes $\alpha_- = f_L(x_-)$.

For convenience we also recall the following result from [CGP+17].

Lemma 6.4.9 ([CGP+17, Lemma 4.1 (i)]) *Introduce the linear operator $S : L_{1,loc}(\mathbb{R}) \to L_{1,loc}(0,\infty)$ defined by*

$$(Sf)(\lambda) = \frac{1}{\pi} \int_0^{\lambda^{1/2}} (\lambda - \nu^2)^{-1/2} f(\nu)d\nu, \quad \lambda > 0.$$

If 0 is a right Lebesgue point for $f \in L_{1,loc}(\mathbb{R})$, then it is also right Lebesgue point for Sf and

$$(Sf)_L(0_+) = \frac{1}{2} f_L(0_+).$$

To establish our results for resolvent regularisations of the Witten index we also need the following lemma, which establishes a more general version of [CGP+17, Lemma 4.1 (ii)]. Its proof follows the argument of [CGP+17, Lemma 4.1 (ii)].

Lemma 6.4.10 *Let $k \in \mathbb{N}$. Introduce the linear operator*

$$T_k : L_1\big((0,\infty); (\nu+1)^{-k-1}d\nu\big) \to L_{1,loc}((0,\infty); d\lambda)$$

by setting

$$(T_k f)(\lambda) = -k\lambda^k \int_0^{\infty} (\nu+\lambda)^{-k-1} f(\nu)d\nu, \quad \lambda > 0.$$

If 0 is a Lebesgue point for $f \in L_1\big((0,\infty); (\nu+1)^{-k-1}d\nu\big)$, then

$$\lim_{\lambda\downarrow 0} (T_k f)(\lambda) = f_L(0_+).$$

Proof Since $T_k\chi_{(0,\infty)} = 1$, we assume, without loss of generality, that 0 is a Lebesgue point for $f \in L^1\big((0,\infty); (\nu+1)^{-k-1}d\nu\big)$ and $f_L(0_+) = 0$. Since $f_L(0_+) = 0$, for a fixed $\delta > 0$ we can find $\tau > 0$ such that

$$\frac{1}{h} \int_0^h |f(\nu)|\, d\nu < \delta, \text{ for all } 0 < h < \tau. \tag{6.16}$$

For every $0 < h < \tau$, we can write

$$-\frac{1}{k}(T_k f)(h) = \int_0^h \frac{h^k f(\nu)\,d\nu}{(\nu+h)^{k+1}} + \int_h^\tau \frac{h^k f(\nu)\,d\nu}{(\nu+h)^{k+1}} + \int_\tau^\infty \frac{hf(\nu)\,d\nu}{(\nu+h)^{k+1}}. \tag{6.17}$$

Since $\frac{h}{\nu+h} \leq 1$ for the first integral we estimate

$$\left| \int_0^h \frac{h^k f(\nu) d\nu}{(\nu+h)^{k+1}} \right| \leq \frac{1}{h} \int_0^h |f(\nu)|\,d\nu \underset{h\downarrow 0}{\longrightarrow} 0.$$

For the third integral, the integrability assumption of f guarantees that

$$\left| \int_\tau^\infty \frac{h^k f(\nu) d\nu}{(\nu+h)^{k+1}} \right| \leq h^k \int_\tau^\infty \frac{|f(\nu)|\,d\nu}{\nu^{k+1}} \underset{h\downarrow 0}{\longrightarrow} 0.$$

Next, we estimate the second term on the right-hand side of (6.17). For brevity, set

$$F(t) = \int_0^t d\nu\,|f(\nu)|, \quad t > 0.$$

Then, integrating by parts, we obtain

$$\left| \int_h^\tau \frac{h^k f(\nu)\,d\nu}{(\nu+h)^{k+1}} \right| \leq \int_h^\tau \frac{h^k |f|(\nu)\,d\nu}{\nu^{k+1}} = \frac{h^k F(\nu)}{\nu^{k+1}}\bigg|_h^\tau + (k+1) \int_h^\tau \frac{h^k F(\nu) d\nu}{\nu^{k+2}}.$$

By (6.16), one concludes that $0 \leq F(\nu) \leq \delta\nu$ for $0 \leq \nu \leq \tau$. Therefore,

$$\left| \int_h^\tau \frac{h^k f(\nu)\,d\nu}{(\nu+h)^{k+1}} \right| \leq \delta + (k+1) h^k \delta \int_h^\tau \frac{d\nu}{\nu^{k+2}} \leq 3\delta.$$

It follows that

$$\limsup_{h\downarrow 0} |(Tf)(h)| \leq 3k\delta.$$

Since δ is arbitrarily small, the assertion follows. □

We are now in a position to state the first main corollary of the principal trace formula. Its proof closely follows the argument used in [CGP⁺17].

Theorem 6.4.11 *Assume Hypothesis 3.2.5 and assume that* 0 *is a right and a left Lebesgue point of* $\xi(\cdot; A_+, A_-)$. *Then* 0 *is a right Lebesgue point of* $\xi(\cdot; \boldsymbol{H}_2, \boldsymbol{H}_1)$

$$\xi_L(0_+; \boldsymbol{H}_2, \boldsymbol{H}_1) = [\xi_L(0_+; A_+, A_-) + \xi_L(0_-; A_+, A_-)]/2$$

and the Witten indices $W_s(\boldsymbol{D}_{\boldsymbol{A}})$ and $W_{k,r}(\boldsymbol{D}_{\boldsymbol{A}})$, $k \geq m$, exist and equal

$$\begin{aligned} W_s(\boldsymbol{D}_{\boldsymbol{A}}) &= W_{k,r}(\boldsymbol{D}_{\boldsymbol{A}}) = \xi_L(0_+; \boldsymbol{H}_2, \boldsymbol{H}_1) \\ &= [\xi_L(0_+; A_+, A_-) + \xi_L(0_-; A_+, A_-)]/2 \end{aligned}$$

Proof First, one rewrites (6.8) in the form,

$$\xi(\lambda; \boldsymbol{H}_2, \boldsymbol{H}_1) = \frac{1}{\pi} \int_0^{\lambda^{1/2}} \frac{d\nu \, [\xi(\nu; A_+, A_-) + \xi(-\nu; A_+, A_-)]}{(\lambda - \nu^2)^{1/2}}, \quad \lambda > 0. \tag{6.18}$$

Define the function $f(\nu) = [\xi(\nu; A_+, A_-) + \xi(-\nu; A_+, A_-)]$. Equality (6.18) implies that

$$\xi(\lambda; \boldsymbol{H}_2, \boldsymbol{H}_1) = (Sf)(\lambda), \quad \lambda > 0,$$

where S is defined in Lemma 6.4.9. By assumption, 0 is a right and a left Lebesgue point of $\xi(\,\cdot\,; A_+, A_-)$, and therefore, 0 is a right Lebesgue point of f. Hence, Lemma 6.4.9 guarantees that 0 is a right Lebesgue point of $\xi(\,\cdot\,; \boldsymbol{H}_2, \boldsymbol{H}_1)$ and

$$\xi_L(0_+; \boldsymbol{H}_2, \boldsymbol{H}_1) = \frac{1}{2} f_L(0_+) = \frac{1}{2}(\xi_L(0_+; A_+, A_-) + \xi_L(0_-; A_+, A_-)).$$

Next, to prove the equality for the semigroup regularised Witten index $W_s(\boldsymbol{D}_{\boldsymbol{A}})$ we introduce the function

$$\Xi(r; \boldsymbol{H}_2, \boldsymbol{H}_1) = \int_0^r \xi(s; \boldsymbol{H}_2, \boldsymbol{H}_1) \, ds, \quad r > 0.$$

By Lipschitz-Krein trace formula (4.39) we have that

$$\begin{aligned} \frac{1}{t} \operatorname{tr}\left(e^{-t\boldsymbol{H}_2} - e^{-t\boldsymbol{H}_1}\right) &= -\int_0^\infty \xi(s; \boldsymbol{H}_2, \boldsymbol{H}_1) \, e^{-ts} \, ds \\ &= -\int_0^\infty e^{-ts} d\, \Xi(s; \boldsymbol{H}_2, \boldsymbol{H}_1). \end{aligned} \tag{6.19}$$

We have already established, that 0 is a right Lebesgue point of $\xi(\,\cdot\,; \boldsymbol{H}_2, \boldsymbol{H}_1)$. Hence, one obtains that

$$\lim_{r \downarrow 0+} \frac{\Xi(r; \boldsymbol{H}_2, \boldsymbol{H}_1)}{r} = \Xi'(0_+; \boldsymbol{H}_2, \boldsymbol{H}_1) = \xi_L(0_+; \boldsymbol{H}_2, \boldsymbol{H}_1)$$

exists. Then, an Abelian theorem for Laplace transforms [Wid41, Theorem 1, p. 181] (with $\gamma = 1$) implies that

$$-\lim_{t\to\infty} \operatorname{tr}\big(e^{-tH_2} - e^{-tH_1}\big) = \lim_{r\downarrow 0+} \frac{\Xi(r; H_2, H_1)}{r} = \xi_L(0_+; H_2, H_1).$$

To prove the equality for the k-th resolvent regularisation $W_{k,r}(D_A)$ we write

$$\begin{aligned} W_{k,r}(D_A) &= \lim_{\lambda\uparrow 0}(-\lambda)^k \operatorname{tr}\big((H_2-\lambda)^{-k} - (H_1-\lambda)^{-k}\big) \\ &= -k \lim_{\lambda\uparrow 0} \int_0^\infty (\nu-\lambda)^{-k-1}\xi(\nu; H_2, H_1)\,d\nu \\ &= \lim_{\lambda\uparrow 0} \big(T_k\xi(\cdot; H_2, H_1)\big)(-\lambda), \end{aligned}$$

where T_k is the operator introduced in Lemma 6.4.10. Since 0 is a right Lebesgue point for

$$\xi(\cdot; H_2, H_1) \in L_1\big((0,\infty); (\nu+1)^{-m-1}d\nu\big) \subset L_1\big((0,\infty); (\nu+1)^{-k-1}d\nu\big), k \geq m,$$

Lemma 6.4.10 implies that

$$W_{k,r}(D_A) = \xi_L(0_+; H_2, H_1) = \frac{1}{2}(\xi_L(0_+; A_+, A_-) + \xi_L(0_-; A_+, A_-)),$$

as required.

□

As a corollary of Theorems 6.4.11 and 5.2.7 we obtain the following extension of the Robbin-Salamon theorem for those operators with some essential spectra outside 0 and without the assumption that the asymptotes $A_\pm$ are boundedly invertible. As we will show in Sect. 7.1 below our framework is suitable for differential operators on non-compact manifolds in any dimension.

Theorem 6.4.12 *Assume Hypothesis 3.2.5 and assume that the asymptotes $A_\pm$ are Fredholm. Then the Witten index of the operator D_A exists and equals*

$$\begin{aligned} W_s(D_A) &= \frac{1}{2}\big(\xi(0_+; A_+, A_-) + \xi(0_-; A_+, A_-)\big) \\ &= \operatorname{sf}\{A(t)\}_{t=-\infty}^{\infty} - \frac{1}{2}[\dim(\ker(A_+)) - \dim(\ker(A_-))]. \end{aligned}$$

6.5 Cyclic Homology and Invariance

We give a brief overview of the homological interpretation of the resolvent form of the principal trace formula following [CK17] and [CGP+17]. Invariance of the Fredholm index under compact perturbations is key to understanding its topological interpretation in terms of K-theory. As the Witten index is not invariant under (relatively) compact perturbations there can be no direct connection to K-theory. Somewhat surprisingly, the Witten index is related instead to cyclic homology, as we now explain by relating the approach of [GS88], the Carey–Pincus point of view in [CP86], and the (unpublished) thesis of Jens Kaad.

In [CP86] bounded non-Fredholm operators T, T^* with the property that the commutator $[T, T^*]$ is trace-class are studied. An important tool exploited there is the Carey–Pincus principal function. It is related directly to the spectral shift function. Kaad's thesis provides us with the connection between [CP86] and cyclic homology and, as we show at the end of this section, with the Witten index.

We start with a brief summary of Chap. 1 of Kaad's thesis: The point of departure for this discussion is a pair of Banach algebras A and J where J is an ideal in A (not necessarily closed). We let $\mathcal{L} = A/J$. By the zeroth relative continuous cyclic homology group of the pair (J, A) we will understand the quotient space $HC_0(J, A) = J/\mathrm{Im}(b)$. Here $b : J \otimes A + A \otimes J \to J$ is an extension of the Hochschild boundary and is given by

$$b : s \otimes a + a' \otimes t \mapsto sa - as + a't - ta',$$

where $s, t \in J$ and $a, a' \in A$. Note that as a topological vector space $HC_0(J, A)$ is non-Hausdorff in general. That is, the image of the extended Hochschild boundary b is not necessarily closed in J. We will also make use of the first cyclic homology group of $\mathcal{L}$ denoted $HC_1(\mathcal{L})$. In the book of Loday [Lod98] he introduces the chain complex $C_m = \mathcal{A}^{\otimes m+1}$, $m = 0, 1, 2, \ldots$, and the two boundary maps b, B being respectively the Hochschild boundary and the Connes' boundary. These are defined on C_m by:

$$\begin{aligned} b(a_0, a_1, \ldots, a_n) &= (a_0 a_1, a_2, \ldots, a_n) + \sum_{i=1}^{n-1} (-1)^i (a_0, a_1, \ldots, a_i a_{i+1}, \ldots, a_n) \\ &\quad + (-1)^n (a_n a_0, a_1, \ldots, a_{n-1}), \end{aligned}$$

with $b : C_m \to C_{m-1}$, and

$$\begin{aligned} B(a_0, \ldots, a_n) &= \sum_{i=0}^{n} (-1)^{ni} (1, a_i, \ldots, a_i, a_0, \ldots, a_{i-1}) \\ &\quad + (-1)^{ni} (a_i, 1, a_{i+1}, \ldots, a_n, a_0, \ldots, a_{i-1}), \end{aligned}$$

where $(a_0, a_1, \ldots, a_n) \in C_m$ and $B : C_m \to C_{m+1}$. Then it is not difficult to check that we have the relations $b^2 = 0 = B^2$, $0 = bB + Bb$. As in [Lod98] we can form a bi-complex $B(\mathcal{A})$ with $B(\mathcal{A})_{p,q} = \mathcal{A}^{\otimes^{q-p+1}}$ with total boundary $b+B$ and hence there are homology groups $\ker(b+B)/\mathrm{Im}(b+B)$ in each degree (being those of periodic cyclic homology) associated with the total boundary $b+B$. We understand $HC_1(\mathcal{L})$ as the first periodic cyclic homology group of $\mathcal{L}$. Let $\mathcal{E}$ be the C^*-algebra generated by T and T^*. Let $J = \mathcal{E} \cap \mathcal{L}_1(\mathcal{H})$. Following Kaad, introduce the exact sequence

$$X : 0 \to J \xrightarrow{i} \mathcal{E} \xrightarrow{q} \mathcal{L} \to 0,$$

where i is the obvious inclusion and q is the quotient map to $\mathcal{L} = \mathcal{E}/J$. Kaad in his thesis proves the following result:

Theorem 6.5.1 *The operator trace determines a well defined map on the zeroth continuous relative cyclic homology group*

$$\mathrm{Tr}_* : HC_0(J, \mathcal{E}) \to \mathbb{C}.$$

This can be extended to a map on $HC_1(\mathcal{L})$ using the connecting map

$$\partial_X : HC_1(\mathcal{L}) \to HC_0(J, \mathcal{E}),$$

coming from the above exact sequence labelled X. Notice that $\mathcal{L}$ is a commutative algebra. The pair $q(T), q(T^*)$ defines a class $q(S) \otimes q(T^*)$ in $HC_1(\mathcal{L})$. This class maps to the commutator $[T, T^*]$ under ∂_X.

6.5.1 How the Witten Index Relates to This

We will adopt a more general viewpoint here than in the earlier parts of these lecture notes. We suppose we have a complex separable Hilbert space $\mathcal{K}$ and form $\mathcal{K}^{(2)} = \mathcal{K} \oplus \mathcal{K}$. We assume we have a linear, closed, densely defined operator S on $\mathcal{K}$ and that S and its adjoint can be combined to form a self-adjoint unbounded operator $\mathcal{D} = \begin{pmatrix} 0 & S^* \\ S & 0 \end{pmatrix}$ on $\mathcal{K}^{(2)}$. We restrict to the case where λ is in the intersection of the resolvent sets of SS^* and S^*S and we make the assumption that

$$-\lambda \,\mathrm{tr}\left((S^*S - \lambda)^{-1} - (SS^* - \lambda)^{-1}\right), \quad \lambda < 0,$$

is finite. Let us make a change of notation and set $\mu^2 = -\lambda$ so that the assumption of the previous equation becomes $(1 + \mu^{-2}S^*S)^{-1} - (1 + \mu^{-2}SS^*)^{-1}$ is trace-class. It is clear from this formulation that we can think of μ as scaling S. This suggests that we make the passage to the bounded picture by writing

$$F_{\mathcal{D}}^{\mu} = \mu^{-1}\mathcal{D}(1+\mu^{-2}\mathcal{D}^2)^{-1/2}.$$

Then

$$1-(F_{\mathcal{D}}^{\mu})^2 = (1+\mu^{-2}\mathcal{D}^2)^{-1} = \begin{pmatrix} (1+\mu^{-2}S^*S)^{-1} & 0 \\ 0 & (1+\mu^{-2}SS^*)^{-1} \end{pmatrix}.$$

For ease of writing let

$$F_{\mathcal{D}}^{\mu} = \begin{pmatrix} 0 & S_{\mu}^* \\ S_{\mu} & 0 \end{pmatrix}.$$

In other words,

$$T_{\mu} = \mu^{-1}S(1+\mu^{-2}S^*S)^{-1/2},$$

and

$$1-(F_{\mathcal{D}}^{\mu})^2 = \begin{pmatrix} 1-T_{\mu}^*T_{\mu} & 0 \\ 0 & 1-T_{\mu}T_{\mu}^* \end{pmatrix}.$$

Consequently, our assumption that $(1+\mu^{-2}S^*S)^{-1}-(1+\mu^{-2}SS^*)^{-1}$ is trace-class translates in the bounded picture to the assumption that the commutator $[T_{\mu}, T_{\mu}^*]$ is trace-class. The Witten index is thus calculating $\lim_{\mu\to 0}\mathrm{tr}_{\mathcal{K}}([T_{\mu}, T_{\mu}^*])$ whenever this limit exists. This puts us into the framework of Kaad's analysis by letting $\mathcal{E}$ be the C^*-algebra generated by T_1 and T_1^* and using the fact that this algebra contains the operators T_{μ}, T_{μ}^* for $\mu > 0$. Thus we see that the theorem quoted above from Kaad's thesis implies that the Witten index is given by a scaling limit of a functional defined on $HC_1(\mathcal{L})$.

6.5.2 Higher Schatten Classes

The generalisation of the Carey-Pincus work that allows its application to differential operators on higher dimensional manifolds begins with a bounded operator T on $\mathcal{H}$ such that

$$(1-T^*T)^n - (1-TT^*)^n$$

is in the trace-class. For $n = 1$ this condition reduces to the trace-class commutator condition. For $n > 1$ we have:

Lemma 6.5.2 *If* $(1 - TT^*)^n - (1 - T^*T)^n$ *is trace-class then* T *and* T^* *commute modulo the* n^{th} *Schatten class.*

Proof This result is a Corollary of [PS10, Theorem 16]. □

The converse to Lemma 6.5.2 appears to require additional side conditions. If the two terms under the trace above are separately trace-class then T is Fredholm and the trace of $(1 - TT^*)^n - (1 - T^*T)^n$ is referred to as the Calderón formula for the Fredholm index of T.

The next step in [CK17] is to introduce certain homology groups of the *-algebra generated by T. This generalises the idea in Kaad's thesis. Within this algebra, denoted by $\mathcal{A}$, there are two ideals $\mathcal{I}$ and $\mathcal{J}$ with $\mathcal{J} \subset \mathcal{I}$. Here $\mathcal{J}$ is chosen to be the ideal generated by $(1 - T^*T)^n - (1 - TT^*)^n$ while $\mathcal{I}$ is the smallest ideal containing $(1 - T^*T)^n$ and $(1 - TT^*)^n$. The key innovation in [CK17] is to introduce a bicomplex for the algebra $\mathcal{A}$ by using the pair of ideals $\mathcal{I}$ and $\mathcal{J}$. The homology theory of the bicomplex of [CK17] has a dual cohomology theory and the pairing between the two, in the concrete situation of operators on Hilbert space, produces the real number $\mathrm{Tr}((1-T^*T)^n-(1-TT^*)^n)$ that is called the homological index in [CK17]. Their version of topological invariance arises from establishing the homotopy invariance of this pairing in the sense of cyclic homology.

One may mimic the scaling argument used for the case $n = 1$ (replace T by T_μ) above to define the Witten index as the scaling limit of the homological index. In this abstract point of view nothing is known about the dependence of the homological index on either μ or n. However it is known that they are connected, n independence of the homological index is equivalent to μ independence.

6.6 The Anomaly in Terms of the Spectral Shift Function

In this section we explain how the spectral shift function $\xi(\cdot; A_+, A_-)$ can be related to the so-called anomaly of the operator $\boldsymbol{D}_{\boldsymbol{A}}$. We begin with a brief discussion of the origin of the term.

6.6.1 The Origin of the Notion of an 'Anomaly'

The term 'anomaly' was coined by physicists in the 1960s to describe surprising behaviour of symmetries in quantum gauge theories. A history with full references to the original papers may be found in the book of Mickelsson [Mic89]. Roughly speaking the problem arises from perturbing the Dirac operator by connections in a suitable affine space and then investigating the invariance of the system under symmetries (in particular, in the terminology of [Mic89], the group of chiral gauge symmetries). As long as one works with classical differential equations there are no problems but if one wants to create quantum field theories based on the classical equations one finds that these chiral symmetries do not carry over. The phenomenon occurs in the very simplest case of two dimensional quantum electrodynamics (where the closely related notion of anomalous commutators was investigated by Schwinger in the 1950s [Sch59]).

The surprising feature that emerges from the calculations performed in studying anomalous symmetry behaviour in these quantum field theories is that the local differential forms involving the characteristic classes in the expression for the index of these Dirac type operators proved by Atiyah-Singer appear. When this first occurred physicists referred to these expressions as anomalies simply because they indicated symmetry breaking. It was only much later that the connection of the quantum field theory calculations to mathematical work in index theory was made.

6.6.2 Relationship to the Spectral Shift Function

Let T be a closed densely defined operator in $\mathcal{H}$. The *anomaly* $\mathrm{Anom}(T)$ of T is define as

$$\mathrm{Anom}(T) = \lim_{t\downarrow 0} \mathrm{tr}\left(e^{-tT^*T} - e^{-tTT^*}\right),$$

whenever this limit exists. Hence, when one wants to compute the anomaly $\mathrm{Anom}(\boldsymbol{D}_{\boldsymbol{A}})$ of the model operator $\boldsymbol{D}_{\boldsymbol{A}}$ in terms of the asymptotes $A_\pm$, one can take the limit of the principal trace formula (see Theorem 6.2.2). As we show below, one can treat this limit in a similar way to the limit taken in the proof of Theorem 6.4.11.

Theorem 6.6.1 *Assume Hypothesis 3.2.5. If the limit*

$$\lim_{t\to\infty} \frac{1}{t^2} \int_{-t}^{t} \sqrt{t^2-u^2}\,\xi(u; A_+, A_-)du$$

exists, then the anomaly $\mathrm{Anom}(\boldsymbol{D}_{\boldsymbol{A}})$ *of the operator* $\boldsymbol{D}_{\boldsymbol{A}}$ *exists and equals*

$$\mathrm{Anom}(\boldsymbol{D}_{\boldsymbol{A}}) = \frac{2}{\pi} \lim_{t\to\infty} \frac{1}{t^2} \int_{-t}^{t} \xi(u; A_+, A_-)\sqrt{t^2-u^2}\,du.$$

Proof We firstly show that the assumption that

$$\lim_{t\to\infty} \frac{1}{t^2} \int_{-t}^{t} \sqrt{t^2-u^2}\,\xi(u; A_+, A_-)du$$

exists, guarantees that the limit $\lim_{t\to\infty} \int_0^t \xi(\lambda; \boldsymbol{H}_2, \boldsymbol{H}_1)d\lambda$ exists too.

Define the function $f(\nu) = [\xi(\nu; A_+, A_-) + \xi(-\nu; A_+, A_-)]$. By Pushnitski's formula (see Theorem 6.3.1) we have

$$\xi(\lambda; \boldsymbol{H}_2, \boldsymbol{H}_1) = \frac{1}{\pi}\int_0^{\lambda^{1/2}} \frac{f(\nu)\,d\nu}{(\lambda - \nu^2)^{1/2}}, \quad \lambda > 0.$$

Therefore, for every fixed $t > 0$, by Fubini's theorem we have

$$\begin{aligned}
\int_0^t \xi(\lambda; \boldsymbol{H}_2, \boldsymbol{H}_1)d\lambda &= \frac{1}{\pi}\int_0^t \Big(\int_0^{\lambda^{1/2}} \frac{f(\nu)\,d\nu}{(\lambda - \nu^2)^{1/2}}\Big)d\lambda \\
&= \frac{1}{\pi}\int_0^{t^{1/2}} \xi(\nu; A_+, A_-)\Big(\int_{\nu^2}^t (\lambda - \nu^2)^{-1/2}d\lambda\Big)d\nu \\
&= \frac{2}{\pi}\int_0^{t^{1/2}} \xi(\nu; A_+, A_-)\sqrt{t - \nu^2}d\nu \\
&= \frac{2}{\pi}\int_{-t^{1/2}}^{t^{1/2}} \xi(\nu; A_+, A_-)\sqrt{t - \nu^2}d\nu.
\end{aligned}$$

Thus, we conclude that

$$\begin{aligned}
\lim_{t\to\infty} \frac{1}{t}\int_0^t \xi(\lambda; \boldsymbol{H}_2, \boldsymbol{H}_1)d\lambda &= \frac{2}{\pi}\lim_{t\to\infty}\frac{1}{t}\int_{-t^{1/2}}^{t^{1/2}} \xi(\nu; A_+, A_-)\sqrt{t - \nu^2}d\nu \\
&= \frac{2}{\pi}\lim_{t\to\infty}\frac{1}{t^2}\int_{-t}^{t} \xi(\nu; A_+, A_-)\sqrt{t^2 - \nu^2}d\nu < \infty,
\end{aligned} \tag{6.20}$$

as required.

Next, as in the proof of Theorem 6.4.11 we introduce the function

$$\Xi(r; \boldsymbol{H}_2, \boldsymbol{H}_1) = \int_0^r \xi(s; \boldsymbol{H}_2, \boldsymbol{H}_1)\,ds, \quad r > 0$$

and using (6.19), we write

$$\frac{1}{t}\mathrm{tr}\big(e^{-t\boldsymbol{H}_2} - e^{-t\boldsymbol{H}_1}\big) = -\int_0^\infty e^{-ts}d\,\Xi(s; \boldsymbol{H}_2, \boldsymbol{H}_1).$$

Then, an Abelian theorem for Laplace transforms [Wid41, Theorem 1, p. 181] (with $\gamma = 1$) implies that

$$-\lim_{t\to 0}\mathrm{tr}\big(e^{-t\boldsymbol{H}_2} - e^{-t\boldsymbol{H}_1}\big) = \lim_{r\uparrow\infty}\frac{\Xi(r; \boldsymbol{H}_2, \boldsymbol{H}_1)}{r} = \lim_{r\to\infty}\frac{1}{r}\int_0^r \xi(s; \boldsymbol{H}_2, \boldsymbol{H}_1)\,ds.$$

Referring to (6.20) we conclude that

$$\mathrm{Anom}(\boldsymbol{D}_{\boldsymbol{A}}) = \lim_{t \to 0} \operatorname{tr}\big(e^{-t\boldsymbol{H}_2} - e^{-t\boldsymbol{H}_1}\big) = \frac{2}{\pi} \lim_{t \to \infty} \frac{1}{t^2} \int_{-t}^{t} \xi(\nu; A_+, A_-)\sqrt{t^2 - \nu^2}\, d\nu,$$

as required. □

Referring to (7.20) we conclude that

$$[illegible]$$

as required. □

Chapter 7
Examples

In this chapter we supplement the abstract discussion by several examples for which our general assumption holds and hence the results of Chap. 6.

Firstly, in Sect. 7.1, we discuss our primary example, the multidimensional Dirac operator and its perturbations given by multiplication operators by matrix valued functions. For a sufficiently good potential the example satisfies the main Hypothesis 3.2.5 and thus, our framework is indeed suitable for differential operators on certain non-compact manifolds. Furthermore, we explain the results in the recent monograph [CGL+22] where theorems on the properties of the spectral shift function for a pair consisting of a Dirac operator and its perturbation by a suitably fast decaying multiplication operator are obtained. We also summarise what is proved there about the Witten index for this case and highlight the exceptional two dimensional case. It is possible that there is an application of the two dimensional case to mathematical models of graphene.

In order to be able to give some concrete computations we study the one-dimensional Dirac operator on the circle (Sect. 7.2) and we compute explicitly the spectral shift function $\xi(\cdot; A_+, A_-)$. This explicit computation allows us to compute both the Witten index of the corresponding operator $\boldsymbol{D}_{\boldsymbol{A}}$ as well as the anomaly. Further one dimensional examples illustrating the effect of the continuous spectrum may be found in our review [CGG+16] and references therein.

7.1 The Dirac Operator in $\mathbb{R}^d$

In this section we present results of the recent monograph [CGL+22]. The main result of [CGL+22] show that the spectral shift function for Dirac operator on $\mathbb{R}^d$, $d \geq 2$, is left and right continuous at zero.

A. Carey, G. Levitina, *Index Theory Beyond the Fredholm Case*, Lecture Notes in Mathematics 2323, https://doi.org/10.1007/978-3-031-19436-8_7

7.1.1 The Setting

Throughout this section we fix $d \in \mathbb{N}, d \geq 2$. For the one dimensional example the reader is referred to [CGLS16b].

For each $k = 1, \dots, d$, we denote by ∂_k the operators of partial differentiation, that is operators in $L_2(\mathbb{R}^d)$ defined as

$$\partial_k = -i\frac{\partial}{\partial t_k}, \quad \mathrm{dom}(\partial_k) = W^{1,2}(\mathbb{R}^d).$$

We denote the tuple $(\partial_1, \dots, \partial_d)$ by ∇.

Let $n(d) = 2^{\lceil \frac{d}{2} \rceil}$. Let $\gamma_k \in M_{n(d)}(\mathbb{C})$, $0 \leq k \leq d$, be Clifford algebra generators, that is,

1. $\gamma_k = \gamma_k^*$ and $\gamma_k^2 = 1$ for $0 \leq k \leq d$.
2. $\gamma_{k_1}\gamma_{k_2} = -\gamma_{k_2}\gamma_{k_1}$ for $0 \leq k_1, k_2 \leq d$, such that $k_1 \neq k_2$.

We use the notation $\gamma = (\gamma_1, \dots, \gamma_d)$.

Definition 7.1.1 Let $m \geq 0$. Define the Dirac operator as an unbounded operator $\mathcal{D}$ acting in the Hilbert space $\mathbb{C}^{n(d)} \otimes L_2(\mathbb{R}^d)$ with domain $\mathrm{dom}(\mathcal{D}) = \mathbb{C}^{n(d)} \otimes W^{1,2}(\mathbb{R}^d)$ by the formula

$$\mathcal{D} = \gamma \cdot \nabla + m\gamma_0. \tag{7.1}$$

It is well-known that the operator $\mathcal{D}$ is self-adjoint. Furthermore,

$$\mathcal{D}^2 = -\Delta + m^2,$$

where $\Delta = \sum_{k=1}^{d} \frac{\partial^2}{\partial^2 x_k}$ is the Laplace operator.

Suppose that $V = \{\varphi_{ij}\}_{i,j=1}^{n(d)}$ is a hermitian matrix of functions, such that $\varphi_{ij} \in L_\infty(\mathbb{R}^d)$. We also write V for the bounded operator of multiplication by this matrix function V on the Hilbert space $\mathbb{C}^{n(d)} \otimes L_2(\mathbb{R}^d)$. We also identify a function $f \in L_\infty(\mathbb{R}^d)$ with the operator on $L_2(\mathbb{R}^d)$ of multiplication by f.

For the example of the current section we set

$$A_- = \mathcal{D}, \quad B_+ = V$$

and

$$B(t) = \theta(t)B_+,$$

where θ satisfies (3.33).

7.1.2 Verification of Hypothesis 3.2.5

In this subsection we gradually impose assumptions on the matrix $V = \{\varphi_{ij}\}$, such that the family $\{\theta(t)V\}$ satisfies Hypothesis 3.2.10 with $p = d$. By Proposition 3.2.11 this ensures that Hypothesis 3.2.5 is also satisfied.

Proposition 7.1.2 *Assume that $V = \{\varphi_{ij}\}_{i,j=1}^{n(d)}$ is such that $\varphi_{ij} \in L_\infty(\mathbb{R}^d)$ and $|\varphi_{ij}(x)| \leq \text{const}\langle x\rangle^{-\rho}$ for some $\rho > d$. Then for any $\varepsilon > 0$ we have that*

$$V(\mathcal{D}^2 + 1)^{-\frac{d}{2}-\varepsilon} \in \mathcal{L}_1(\mathcal{H}).$$

In particular, V is a d-relative trace-class perturbation with respect to $\mathcal{D}$, that is Hypothesis 3.2.10 (ii) is satisfied.

Proof Cwikel estimates in the trace-class ideal (see e.g. [Sim05, Theorem 4.4], [LSZ20]) imply that $\langle\cdot\rangle^{-\rho}(-\Delta + 1)^{-\frac{d}{2}-\varepsilon}$ is a trace-class operator. The assumption on V together with the equality

$$V(\mathcal{D}^2 + 1)^{\frac{-d}{2}-\varepsilon} = \left(\varphi_{ij}(-\Delta + 1)^{\frac{-d}{2}-\varepsilon}\right)_{i,j=1}^{n(d)}$$

imply that is $V(\mathcal{D}^2 + 1)^{\frac{-d}{2}-\varepsilon}$ a trace-class operator (on $\mathbb{C}^{n(d)} \otimes L_2(\mathbb{R}^d)$). □

We recall that the operator $L^k_{\mathcal{D}^2}$, $k \in \mathbb{N}$, is defined in (3.35).

Proposition 7.1.3 *Let $k \in \mathbb{N}$ be fixed. Assume that $V = \{\varphi_{ij}\}_{i,j=1}^{n(d)}$ is such that $\varphi_{ij} \in W^{2k,\infty}(\mathbb{R}^n)$, $i, j = 1, \dots n(d)$. Then $V \in \bigcap_{j=1}^k \text{dom}(L^j_{\mathcal{D}^2})$.*

Proof Let $k \in \mathbb{N}$ be fixed. We firstly note that

$$L^k_{\mathcal{D}^2}(V) = \left(L^k_{-\Delta}(\varphi_{ij})\right)_{i,j=1}^{n(d)},$$

and therefore, it is siffucient to show that $\varphi \in \bigcap_{j=1}^k \text{dom}(L^j_{-\Delta})$ for any $\varphi \in W^{2k,\infty}(\mathbb{R}^n)$.

Since $\varphi \in W^{2k,\infty}(\mathbb{R}^n)$, we have that $\varphi\,\text{dom}(\Delta)^j \subset \text{dom}(\Delta)^j$ for all $j = 1, \dots, 2k$. Furthermore, since

$$[\partial_k, \varphi]\xi = \frac{1}{i}\frac{\partial\varphi}{\partial x_k}\xi, \quad k = 1, \dots, d, \quad \varphi \in W^{1,\infty}(\mathbb{R}^n),$$

we have

$$\begin{aligned}
[\Delta,\varphi] &= \sum_{j=1}^{n}[\partial_j^2,\varphi] = \sum_{j=1}^{n}\partial_j[\partial_j,\varphi] + \sum_{j=1}^{n}[\partial_j,\varphi]\partial_j \\
&= \frac{1}{i}\sum_{j=1}^{n}\Big(\partial_j\frac{\partial\varphi}{\partial x_j} + \frac{\partial\varphi}{\partial x_j}\partial_j\Big) = \frac{1}{i}\sum_{j=1}^{n}\Big(2\partial_j\frac{\partial\varphi}{\partial x_j} - [\partial_j,\frac{\partial\varphi}{\partial x_j}]\Big) \\
&= \frac{2}{i}\sum_{j=1}^{n}\partial_j\frac{\partial\varphi}{\partial x_j} + \sum_{j,\ell=1}^{n}\frac{\partial^2\varphi}{\partial x_j\partial x_\ell}.
\end{aligned}$$

Therefore,

$$(1-\Delta)^{-1/2}[\Delta,\varphi] = \frac{2}{i}\sum_{j=1}^{n}\partial_j(1-\Delta)^{-1/2}\frac{\partial\varphi}{\partial x_j} + \sum_{j,\ell=1}^{n}(1-\Delta)^{-1/2}\frac{\partial^2\varphi}{\partial x_j\partial x_\ell}.$$

Since the operator $\partial_j(1-\Delta)^{-1/2}$ is bounded, we infer that

$$\overline{(1-\Delta)^{-1/2}[\Delta,\varphi]} \in \mathcal{L}(L_2(\mathbb{R}^n)).$$

Continuing this process, we obtain that

$$\overline{(1-\Delta)^{-j}[\Delta,\varphi]^{(j)}} \in \mathcal{L}(L_2(\mathbb{R}^n)), \quad j=1,\dots,k,$$

that is $\varphi \in \bigcap_{j=1}^{2k}\operatorname{dom}(L_{-\Delta}^{j})$. □

We now state the assumptions on the potential V which ensure that the result of Chap. 6 applicable to the Dirac operator on $\mathbb{R}^d$.

Hypothesis 7.1.4 *Assume that* $V = \{\varphi_{ij}\}_{i,j=1}^{n(d)}$ *is such that*

$$\varphi_{ij} \in W^{4d,\infty}(\mathbb{R}^d),\ |\varphi_{ij}(x)| \le \langle x\rangle^{-\rho}, \quad i,j=1,\dots,n(d)$$

for some $\rho > d$.

Combining now Propositions 7.1.2 and 7.1.3 we arrive at the following

Theorem 7.1.5 *Let* $\mathcal{D}$ *be the Dirac operator on* $C^{n(d)} \otimes L_2(\mathbb{R}^d)$ *defined by* (7.1), $d \in \mathbb{N}$ *and let a potential* V *satisfy Hypothesis 7.1.4. Then the operator* $A_- = \mathcal{D}$ *and the perturbation* $B_+ = V$ *satisfy Hypothesis 3.2.10 (and hence also Hypothesis 3.2.5) with* $p = d$.

7.1.3 *The Index of* $\boldsymbol{D_A}$

Everywhere below we assume that the perturbation $V = \{\varphi_{ij}\}_{i,j=1}^{n(d)}$ satisfies Hypothesis 7.1.4. Since these assumptions guarantee that Hypothesis 3.2.10 is satisfied for the Dirac operator, Theorem 6.4.11 implies that the Witten index can be expressed via the spectral shift function $\xi(\cdot; \mathcal{D}+V, \mathcal{D})$, provided that the latter one is well-behaved at zero.

Since $\mathcal{D}^2 = -\Delta + m^2$, it follows that the operator $\mathcal{D}$ has purely absolutely continuous spectrum, which coincides with $(-\infty - m] \cup [m, \infty)$. In the case, when m is strictly positive, the assumption that the operator $V(\mathcal{D}+i)^{-d-1}$ is compact together with Weyl's theorem guarantees that the operator $\mathcal{D}+V$ has purely discrete spectrum in the interval $(-m, m)$. In particular, if $\mathcal{D}+V$ is also invertible, then by Theorem 3.2.3 the corresponding operator

$$\boldsymbol{D}_{\boldsymbol{A}} = \frac{d}{dt} \otimes 1 + 1 \otimes \mathcal{D} + \theta V, \tag{7.2}$$

(see (3.26)) is Fredholm. Furthermore, since in this case the spectral shift function $\xi(\cdot; \mathcal{D}+V, \mathcal{D})$ for the pair $(\mathcal{D}+V, \mathcal{D})$ is constant in a neighbourhood of zero Theorem 6.4.11 implies that

$$\operatorname{index}(\boldsymbol{D}_{\boldsymbol{A}}) = \xi(0; \mathcal{D}+V, \mathcal{D}).$$

If the operator $\mathcal{D}+V$ is not invertible, then the operator $\boldsymbol{D}_{\boldsymbol{A}}$ is no longer Fredholm. However, since in this case the spectral shift function $\xi(\cdot; \mathcal{D}+V, \mathcal{D})$ is left and right continuous at zero, it follows that 0 is, in particular, left and right Lebesgue point of $\xi(\cdot; \mathcal{D}+V, \mathcal{D})$. Hence, Theorem 6.4.11 again implies that

$$W_s(\boldsymbol{D}_{\boldsymbol{A}}) = \frac{1}{2}[\xi(0_+; \mathcal{D}+V, \mathcal{D}) + \xi(0_-; \mathcal{D}+V, \mathcal{D})]. \tag{7.3}$$

In particular, if $d = 3$ and the potential V is a magnetic potential, that is

$$V = \sum_{n=1}^{3} \gamma_j A_j, \quad A_j \in L_3(\mathbb{R}^3) \cap L_1(\mathbb{R}^3),$$

then [Saf01, Section 5.1] implies that $\xi(0; \mathcal{D}+V, \mathcal{D}) = 0$, and therefore, by (7.3) we obtain that

$$W_s(\boldsymbol{D}_{\boldsymbol{A}}) = 0.$$

However, our main interest lies in the massless Dirac operator, where $m = 0$. In this case, the spectrum of $\mathcal{D}$ covers the whole real line, and therefore, whatever the potential V, the operator $\boldsymbol{D}_{\boldsymbol{A}}$ in (7.2) is never Fredholm. To study whether the

Witten index of $\boldsymbol{D}_A$ exists in this case it is sufficient to study the spectral shift function $\xi(\cdot; \mathcal{D}+V, \mathcal{D})$ for the pair $(\mathcal{D}+V, \mathcal{D})$ and its behaviour near zero.

Although the spectral shift function for second order operators is well studied, there is a sparse literature available for the spectral shift function for the first order operators $\mathcal{D}$ and $\mathcal{D}+V$ (see e.g. [Saf01, TdA11, BR99]). Furthermore, the majority of papers on this topic study the massive case, $m > 0$, and are, therefore, not applicable to our case. In the next section we discuss recent results of [CGL+22] which show that for a sufficiently good perturbation V, the spectral shift function $\xi(\cdot; \mathcal{D}+V, \mathcal{D})$ is left and right continuous at zero and hence the Witten index of the operator $\boldsymbol{D}_A$ exists.

7.1.4 Behaviour of the Spectral Shift Function for the Massless Dirac Operator

In this section we summarise those parts of [CGL+22] that are directly relevant to the earlier material in these notes.

The main restriction we place on V is the following:

Hypothesis 7.1.6 *Let $d \in \mathbb{N}$ and suppose that $V = \{\varphi_{ij}\}_{1\le i,j\le n(d)}$ is self-adjoint and satisfies for some constants $C \in (0,\infty)$ and $\varepsilon > 0$,*

$$\varphi_{ij} \in L_\infty(\mathbb{R}^d), \quad |\varphi_{ij}(x)| \le C\langle x\rangle^{-d-1-\varepsilon} \text{ for a.e. } x \in \mathbb{R}^d, \ 1 \le i, j \le n(d).$$

Proposition 7.1.2 and Theorem 4.2.12 guarantee that we can choose spectral shift function $\xi(\cdot; \mathcal{D}+V, \mathcal{D})$ such that $\xi(\cdot; \mathcal{D}+V, \mathcal{D}) \in L_1\big(\mathbb{R}; (1+|\lambda|)^{-d-1}\,d\lambda\big)$. By Remark 4.2.13 this spectral shift function may differ by an additive constant from the spectral shift function fixed via (4.31). However, as we are interested in the a.e. pointwise behaviour of $\xi(\cdot; \mathcal{D}+V, \mathcal{D})$ this additive constant is irrelevant for the discussion of the present section.

Furthermore, by Theorem 4.4.4 for a.e. $\lambda \in \mathbb{R}$ we have

$$\begin{aligned}\xi(\lambda; \mathcal{D}+V, \mathcal{D}) &= \pi^{-1}\operatorname{Im}\ln \det_{d+1}(1+V(\mathcal{D}-(\lambda+i0))^{-1}) \\ &\quad - \pi^{-1}\operatorname{Im}(G_{\mathcal{D}+V,\mathcal{D}}(\lambda+i0)) + P_{d-1}(\lambda),\end{aligned} \tag{7.4}$$

provided that the analytic (in $\mathbb{C}\setminus\mathbb{R}$) functions $\ln\det_{d+1}(1+V(\mathcal{D}-\cdot)^{-1})$ and $G_{\mathcal{D}+V,\mathcal{D}}$ have normal boundary values. Here, the function $G_{\mathcal{D}+V,\mathcal{D}}$ is defined in (4.34) and P_{d-1} is a polynomial of degree less than of equal to $d-1$.

The function $G_{\mathcal{D}+V,\mathcal{D}}$ is auxiliary, which only complicates the computation and does not carry any specific information about the spectrum of $\mathcal{D}+V$. In fact we have the following

Theorem 7.1.7 *[CGL+22, Theorem 11.2] Assume Hypothesis 7.1.6. Then the function $G_{\mathcal{D}+V,\mathcal{D}}$ is continuous in $\overline{\mathbb{C}_+}$.*

In particular, the above theorem together with (7.4) implies that the behaviour of the spectral shift function $\xi(\cdot; \mathcal{D}+V, \mathcal{D})$ is determined by the behaviour of the function

$$\ln(\det_{d+1}(1 + V(\mathcal{D} - \cdot)^{-1}))$$

at the boundary.

Using the polar decomposition for V we write

$$V = V_1 V_2, \quad V_1 = |V|^{1/2}, V_2 = U|V|^{1/2}. \tag{7.5}$$

Using this factorisation and elementary properties of regularized determinants, the analysis of the function $\ln(\det_{d+1}(1 + V(\mathcal{D} - \cdot)^{-1}))$ reduces to an analysis of

$$\ln\big(\det_{d+1}\big(1 + V_2(\mathcal{D} - z)^{-1} V_1^*\big)\big), \quad z \in \mathbb{C}_+.$$

Here the operator

$$V_2(\mathcal{D} - z)^{-1} V_1, \quad z \in \mathbb{C}_+ \tag{7.6}$$

is the (symmetrized) Birman–Schwinger operator.

As the following theorem shows, the Birman–Schwinger operator (7.6) defines a $\mathcal{L}_{d+1}(\mathcal{H})$-valued function which is continuous in the closed-upper half-plane.

Theorem 7.1.8 *[CGL⁺22, Theorem 6.16]Assume Hypothesis 7.1.6 and let* $V = V_1^* V_2$ *be the decomposition as in* (7.5)*. Then*

$$V_2(\mathcal{D} - z)^{-1} V_1 \in \mathcal{L}_{d+1}(\mathcal{H}), \quad z \in \overline{\mathbb{C}_+}$$

and $V_2(\mathcal{D} - \cdot)^{-1} V_1$ *is continuous on* $\overline{\mathbb{C}_+}$ *with respect to* $\mathcal{L}_{d+1}(\mathcal{H})$*-norm.*

In particular, Theorem 7.1.8 implies that the boundary values of the regularized Fredholm determinant,

$$\det_{d+1}\big(1 + V_2(\mathcal{D} - (\lambda + i0))^{-1} V_1^*\big),$$

exist and are continuous for all $\lambda \in \mathbb{R}$. This means that the function $\ln(\det_{d+1}(1 + V(\mathcal{D} - \cdot)^{-1}))$ has normal boundary values and is continuous at any point λ in $\mathbb{R}$ such that

$$\det_{d+1}\big(1 + V_2(\mathcal{D} - (\lambda + i0))^{-1} V_1^*\big) \neq 0. \tag{7.7}$$

The classical theory of Kato-smooth operators in scattering theory enables us to establish equality (7.7) for all $\lambda \neq 0$. Namely, as showed in [CGL⁺22, Theorem 3.4] a non-zero number $\lambda \in \mathbb{R}$ is an eigenvalue of $\mathcal{D} + V$ if and only if -1 is an eigenvalue of $1 + V_2(H_0 - (\lambda + i0))^{-1} V_1^*$ (with equal geometric multiplicities).

Equivalently, Eq. (7.7) holds for $\lambda \in \mathbb{R} \setminus \{0\}$ if and only if λ is an eigenvalue of $\mathcal{D}+V$.

The recent result of [KOY15] gives sufficient conditions to exclude the existence of eigenvalues of $\mathcal{D} + V$. In particular, [KOY15, Theorem 2.1,2.3] imply the following

Corollary 7.1.9 *Assume Hypothesis 7.1.6 and assume, in addition, that for some* $R > 0$,

$$V \in \left[C^1(E_R)\right]^{n(d)\times n(d)}, \text{ where } E_R = \{x \in \mathbb{R}^d \mid |x| > R\}, \tag{7.8}$$

and

$$(x \cdot \nabla\varphi_{ij})(x) \underset{|x|\to\infty}{=} o(1), \quad 1 \leq i, j \leq n(d), \quad \text{uniformly with respect to directions.} \tag{7.9}$$

Then, $\sigma_p(\mathcal{D}+V) \subseteq \{0\}$.

Thus, under the assumptions of Corollary 7.1.9 we have that

$$\det{}_{d+1}\left(1 + V_2(\mathcal{D} - (\lambda + i0))^{-1} V_1^*\right) \neq 0, \quad \lambda \neq 0.$$

In particular, under these assumptions the function $\operatorname{Im}(\ln \det_{d+1}\left(1 + V_2(\mathcal{D} - (\lambda + i0))^{-1} V_1^*\right))$ is continuous for $\lambda \in \mathbb{R}\setminus\{0\}$.

The analysis of the function $\ln \det_{d+1}\left(1 + V_2(\mathcal{D} - (\cdot + i0))^{-1} V_1^*\right)$ at zero is significantly more intricate and is related to the existence of a zero eigenvalue or resonance.

We say that 0 is a *zero-energy resonance* of $\mathcal{D} + V$ if $\ker\left(1 + V_2(\mathcal{D} - (\cdot + i0))^{-1} V_1^*\right) \neq \{0\}$ and there exists ψ (which can be constructed from a an eigenvector of $V_2(\mathcal{D} - (\cdot + i0))^{-1} V_1^*$ corresponding to eigenvalue -1), such that ψ is a distributional solution of $(\mathcal{D}+V)u = 0$, satisfying $\psi \notin [L_2(\mathbb{R}^d)]^{n(d)}$. (To avoid confusion about this definition of resonance the reader should consult the theorem below for further constraints on ψ that follow from the use of the Birman-Schwinger operator.) We say that 0 is a *regular point* of $\mathcal{D} + V$ if it has no kernel and nor is there a resonance at zero.

Starting from an element of the kernel of

$$1 + V_2(\mathcal{D} - (0 + i0))^{-1} V_1$$

there is a direct method for producing a corresponding solution of the differential equation $(\mathcal{D} + V)\psi = 0$. While the point 0 being regular for $\mathcal{D} + V$ is the generic situation, zero-energy eigenvalues and/or resonances are exceptional cases.

In fact, we have the following

Proposition 7.1.10 ([CGL⁺22, Theorem 10.7 (iii)]) *Assume Hypothesis 7.1.6. The point 0 is a regular point of $\mathcal{D} + V$ if and only if*

$$\ker\left(1 + V_2(\mathcal{D} - (0 + i0))^{-1} V_1^*\right) \neq \{0\}.$$

In particular, if 0 is a regular point of $\mathcal{D} + V$, then the function $\mathrm{Im}(\ln \det_{d+1}(1 + V_2(\mathcal{D} - (\lambda + i0))^{-1} V_1^*))$ is continuous at $\lambda = 0$, and so is continuous everywhere on the real line.

In [CGL⁺22, Theorem 10.7] under the preceding hypothesis on V we find that the existence of a kernel for $\mathcal{D} + V$ cannot be ruled out in any dimension $n \geq 2$ though resonances are only possible in dimension 2. This fact may be of interest for mathematical models of graphene which use the massless Dirac operator and perturbations as Hamiltonians. Namely, we have

Theorem 7.1.11 ([CGL⁺22, Theorem 10.7]) *Assume Hypothesis 7.1.6.*

(i) If $d = 2$, there are precisely four possible cases:
Case (I): 0 is regular for $\mathcal{D} + V$.
Case (II): 0 is a (possibly degenerate) resonance of $\mathcal{D} + V$. In this case the resonance functions ψ satisfy

$$\psi \in [L_q(\mathbb{R}^2)]^2, \quad q \in (2, \infty) \cup \{\infty\}, \quad \nabla\psi \in [L_2(\mathbb{R}^2)]^{2\times 2},$$
$$\psi \notin [L_2(\mathbb{R}^2)]^2.$$

Case (III): 0 is a (possibly degenerate) eigenvalue of $\mathcal{D} + V$. In this case the corresponding eigenfunctions $\psi \in \mathrm{dom}(\mathcal{D} + V) = \left[W^{1,2}(\mathbb{R}^2)\right]^2$ of $(\mathcal{D} + V)\psi = 0$ also satisfy

$$\psi \in [L_q(\mathbb{R}^2)]^2, \quad q \in [2, \infty) \cup \{\infty\}.$$

Case (IV): A possible mixture of Cases (II) and (III).

(ii) If $d \in \mathbb{N}$, $d \geq 3$, there are precisely two possible cases:
Case (I): 0 is regular for $\mathcal{D} + V$.
Case (II): 0 is a (possibly degenerate) eigenvalue of $\mathcal{D} + V$. In this case, the corresponding eigenfunctions $\psi \in \mathrm{dom}(\mathcal{D} + V) = \left[W^{1,2}(\mathbb{R}^d)\right]^{n(d)}$ of $(\mathcal{D} + V)\psi = 0$ also satisfy

$$\psi \in \left[L_q(\mathbb{R}^d)\right]^{n(d)}, \quad q \in \begin{cases} (3/2, \infty) \cup \{\infty\}, d = 3, \\ (4/3, 4), & d = 4, \\ (2d/(d+2), 2d/(d-2)), & d \geq 5. \end{cases}$$

In particular, there are no zero-energy resonances of $\mathcal{D} + V$ in dimension $d \geq 3$.

Remark 7.1.12 The absence of zero-energy resonances is well-known in the three-dimensional case $n = 3$, see [Aib16], [BE11, Sect. 4.4], [BES08], [BGW95], [SU08b], [SU08a], and [ZG13]. In fact, for $d = 3$ the absence of zero-energy resonances has been shown under the weaker decay $|\varphi_{ij}| \le C\langle x\rangle^{-1-\varepsilon}$, $x \in \mathbb{R}^3$, in [Aib16]. ◇

In [KOY15] a sufficient condition to exclude a zero-energy eigenvalue is given.

Proposition 7.1.13 *Assume that V is as in Corollary 7.1.9 and assume also that*

$$\sup_{x\in\mathbb{R}^n} |x|\|V(x)\|_{\mathcal{L}(\mathbb{C}^N)} \le C \text{ for some } C \in (0, (d-1)/2). \tag{7.10}$$

Then 0 is not an eigenvalue of $\mathcal{D} + V$.

Another sufficient condition to exclude the existence of any eigenvalues of $\mathcal{D} + V$ is given in [CGK+18] for $\|V\|_\infty$ sufficiently small.

At the present moment there is no way to exclude the existence of a zero-energy resonance of $\mathcal{D} + V$ in the two-dimensional case. Nevertheless, using the approach of Jensen and Nenciu [JN01] the asymptotic expansion of

$$\ln(\det_{d+1}(1 + V(\mathcal{D} - z)^{-1})), \quad z \to 0, z \in \overline{\mathbb{C}_+} \setminus 0$$

has been established in [CGL+22]. This asymptotic expansion allowed the authors of [CGL+22] to establish the following result.

Theorem 7.1.14 ([CGL+22, Theorem 10.17]) *Assume Hypothesis 7.1.6. In addition, we assume that V satisfies* (7.8) *and* (7.9). *Then*

$$\ln(\det_{d+1}(1 + V(\mathcal{D} - (\cdot))^{-1})), \quad z \in \mathbb{C}_\pm,$$

has normal boundary values on $\mathbb{R}\setminus\{0\}$ and the boundary values to $\mathbb{R}$ are continuous on $(-\infty, 0) \cup (0, \infty)$ and left and right continuous at zero.

In particular, if 0 is a regular point for $\mathcal{D} + V$ the boundary values of

$$\ln(\det_{d+1}(1 + V(\mathcal{D} - \cdot)^{-1}))$$

are continuous everywhere on the real line.

Combining Theorems 7.1.14 and 7.1.7 with the representation (7.4) the principal result of [CGL+22] is the following.

Theorem 7.1.15 *Let V be as in Theorem 7.1.14. Then and the left and right limits at zero,*

$$\xi(\cdot; \mathcal{D} + V, \mathcal{D}) = \lim_{\varepsilon\downarrow 0} \xi(\pm\varepsilon; \mathcal{D} + V, \mathcal{D}),$$

exist. In particular, if 0 *is a regular point for* $\mathcal{D}+V$

$$\xi(\cdot;\mathcal{D}+V,\mathcal{D})\in C(\mathbb{R}).$$

Remark 7.1.16 By Theorem 7.1.11 there are no zero-energy resonance of $\mathcal{D}+V$ whenever $d\geq 3$. In addition, if V satisfies (7.10), then there are no zero-energy eigenvalues of $\mathcal{D}+V$, so that 0 is a regular point of $\mathcal{D}+V$. Thus, under these assumptions on V, the spectral shift function $\xi(\cdot;\mathcal{D}+V,\mathcal{D})$ is continuous at zero. ◇

A combination of Theorem 7.1.15 and Theorem 6.4.11 implies the following result, which is the first of this kind applicable to non-Fredholm operators in a partial differential operator setting involving massless Dirac operators in all dimensions. It characterizes the Witten index of $\boldsymbol{D_A}$ in terms of spectral shift functions.

Hypothesis 7.1.17 *Let* $d\in\mathbb{N}$ *and suppose that* $V=\{\varphi_{ij}\}_{1\leq i,j\leq n(d)}$ *is self-adjoint and satisfies for some constants* $C\in(0,\infty)$ *and* $\varepsilon>0$,

$$\varphi_{ij}\in W^{4d,\infty}(\mathbb{R}^d),\quad |\varphi_{ij}(x)|\leq C\langle x\rangle^{-d-1-\varepsilon}\ \textit{for a.e.}\ x\in\mathbb{R}^d,\ 1\leq i,j\leq n(d).$$

In addition, we assume that V *satisfies* (7.8) *and* (7.9).

Theorem 7.1.18 *Assume Hypothesis 7.1.17. Then the Witten indices* $W_s(\boldsymbol{D_A})$ *and* $W_{k,r}(\boldsymbol{D_A})$, $k\geq d$, *of the operator* $\boldsymbol{D_A}=\frac{d}{dt}\otimes 1+1\otimes\mathcal{D}+\theta V$ *exist and equal*

$$W_s(\boldsymbol{D_A})=W_{k,r}(\boldsymbol{D_A})==[\xi(0_+;\mathcal{D}+V,\mathcal{D})+\xi(0_-;\mathcal{D}+V,\mathcal{D})]/2.$$

It is well known that if, in the setting of the Robbin-Salamon theorem, one or both end-points $A_\pm$ of the path along which one is calculating spectral flow is not invertible then the model operator is not Fredholm. In the approach described here we have already noted that the Witten index exists in this situation. In the next section we give in detail an example that illustrates the range of phenomena that can occur in this situation.

7.2 A Compact One-Dimensional Example

In this section we study in full detail the case of a one-dimensional operator on a finite interval $[a,b]$ with twisted periodic boundary conditions. This example is, of course, of a path $\{A(t)\}_{t\in\mathbb{R}}$ of operators with purely discrete spectrum. However, the importance of this example is that we do not exclude the cases, when the asymptotes $A_\pm$ of $\{A(t)\}_{t\in\mathbb{R}}$ are not invertible. Our objective is to explain an example in which more or less everything is explicitly computable.

Recall that the assumption that $A_\pm$ are not invertible makes [RS95] inapplicable. For this example we will now compute explicitly the spectral shift function $\xi(\cdot;A_+,A_-)$. This in turn allows us to compute the Witten index of $\boldsymbol{D_A}$, the spectral flow along the path $\{A(t)\}_{t\in\mathbb{R}}$, as well as the anomaly of the operator $\boldsymbol{D}_A$. In particular, we show that, in general, these three numbers are pairwise distinct, a fact that we found somewhat surprising.

7.2.1 The Setting

We introduce $A_- = -i\frac{d}{dx}$ on $L_2[0, 2\pi]$ with twisted periodic boundary conditions and its perturbation B_+ given by a multiplication operator. Specifically, let $\alpha \in [0, 1)$ and let $D_\alpha = \frac{d}{idx}$ be the differentiation operator on $L_2[0, 2\pi]$ with boundary conditions:

$$\mathrm{dom}(D_\alpha) = \{\xi \in L_2[0, 2\pi] : \xi \in AC[0, 2\pi], \quad \xi(0) = e^{i2\pi\alpha}\xi(2\pi)\},$$

where $AC[0, 2\pi]$ denotes the space of all absolutely continuous functions on $[0, 2\pi]$.

Recall that the operator D_α has eigenvalues $\lambda_{\alpha,n}$ and eigenfunctions $e_{\alpha,n}$ given by

$$\lambda_{\alpha,n} = n - \alpha, \quad e_{\alpha,n}(t) = (2\pi)^{-1/2}e^{i\lambda_{\alpha,n}t}, \quad n \in \mathbb{Z}, \tag{7.11}$$

where every eigenvalue has multiplicity 1 (see e.g. [Sto90, Section X.2]). Therefore, for any bounded perturbation $f \in L_\infty[0, 2\pi]$, we have $f(D_\alpha + i)^{-2} \in \mathcal{L}_1(L_2[0, 2\pi])$, which means that the multiplication operator f is 1-relative trace-class perturbation of D_α. (Here, as above, we identify $f \in L_\infty[0, 2\pi]$ with bounded multiplication operator by f on $L_2[0, 2\pi]$).

Recall also (see e.g. [Sto90, Section X.2]), that the operator D_α is unitarily equivalent to the operator $D_0 - \alpha$, namely, introduce the unitary U_α given by multiplication by the function $s \mapsto e^{-i\alpha s}$, $s \in [0, 2\pi]$ and then

$$U_\alpha^* D_\alpha U_\alpha = D_0 - \alpha, \tag{7.12}$$

In contrast to the non-compact one-dimensional example (see [CGG+16, Remark 2.3]), the operators D_α and $D_\alpha + \varphi$ are unitarily equivalent only under an additional condition on φ.

Lemma 7.2.1 *If $\int_0^{2\pi} h(s)ds \in 2\pi\mathbb{Z}$, then the operators $D_\alpha + M_h$ and D_α are unitarily equivalent. That is, for the function*

$$\psi(t) = \exp(-i\int_0^t h(s)ds), \quad t \in [0, 2\pi],$$

we have $\psi D_\alpha \bar{\psi} = D_\alpha + h$.

Proof We firstly note that since $\int_0^{2\pi} h(s)ds = 2\pi\mathbb{Z}$, we have $\psi(0) = \psi(2\pi)$. Since, in addition, $\psi' = -i\psi h$, we have that $\psi\xi \in \mathrm{dom}(D_\alpha)$ for all $\xi \in \mathrm{dom}(D_\alpha)$. Now, for an arbitrary $\xi \in \mathrm{dom}(D_\alpha)$ we have,

$$[\psi, D_\alpha]\xi = \psi D_\alpha \xi - D_\alpha(\psi\xi) = \frac{1}{i}(\psi\xi' - (\xi\psi)') = \frac{-\psi'}{i}\xi = \psi h\xi.$$

Thus, $\psi D_\alpha - D_\alpha \psi = h\psi$ or, equivalently, $(D_\alpha + h)\psi = \psi D_\alpha$ and the claim follows. □

Let $\varphi \in C^1[0, 2\pi]$ be an arbitrary function. We set

$$c = \frac{1}{2\pi}\int_0^{2\pi} \varphi(s)ds \quad and \quad \varphi_0 := \varphi - c \tag{7.13}$$

so that $\int_0^{2\pi} \varphi_0(s)ds = 0$.

Remark 7.2.2 It follows from Lemma 7.2.1 and (7.11) that for any $\varphi \in L_\infty[0, 2\pi]$ we have

$$\begin{aligned}\sigma(D_\alpha + \varphi) &= \sigma(D_\alpha + \varphi_0 + c) = \sigma(D_\alpha + c) = \{\lambda_{\alpha,n} + c, n \in \mathbb{Z}\} \\ &= \Big\{n - \alpha + c, n \in \mathbb{Z}\Big\},\end{aligned}$$

where every eigenvalue has multiplicity one. ◇

Remark 7.2.3 Of course, a knowledge of the spectra of D_α and $D_\alpha + \varphi$ is sufficient to give the jumps of the spectral shift function $\xi(\cdot; D_\alpha + \varphi, D_\alpha)$. However, to compute the Witten index, the spectral flow, as well as the anomaly, the additive constant in $\xi(\cdot; D_\alpha + \varphi, D_\alpha)$ has to be computed too. ◇

7.2.2 *Spectral Shift Function of the Pair* $(D_\alpha + \varphi, D_\alpha)$

Recall that (see (4.41))

$$\mathrm{tr}\,(f(A_+) - f(A_-)) = \int_{\mathbb{R}} f'(\lambda)\cdot\xi(\lambda; A_+, A_-)\,d\lambda, \quad f \in \mathfrak{F}_1(\mathbb{R}),$$

and the spectral shift function $\xi(\cdot; A_+, A_-)$ is fixed by (4.31). Our approach to computing the spectral shift function requires computation of the trace of $f(A_+) - f(A_-)$ for a class of functions f significantly larger than $\mathfrak{F}_1(\mathbb{R})$. The purpose of this section is to show that under some additional conditions, for this specific choice of the spectral shift function $\xi(\cdot; A_+, A_-)$ in (4.31), the trace formula above can be extended to the required class.

Proposition 7.2.4 *Let $A_\pm$, P_n and $A_{+,n}$, $n \in \mathbb{N}$, be as before (see (3.20), (3.5) and (3.24)). Assume that $g(A_+) - g(A_-) \in \mathcal{L}_1(\mathcal{H})$ and*

$$\lim_{n\to\infty} \left\| [g(A_{+,n}) - g(A_-)] - [g(A_+) - g(A_-)] \right\|_1 = 0.$$

Then for any $F \in C_b^\infty(\mathbb{R})$ such that $|F'(\lambda)| \leq \text{const}(1+\lambda^2)^{-1}$, $\lambda \in \mathbb{R}$, we have $F(A_+) - F(A_-) \in \mathcal{L}_1(\mathcal{H})$ and

$$\text{tr}\left(F(A_+) - F(A_-)\right) = \int_{\mathbb{R}} F'(\lambda)\xi(\lambda; A_+, A_-)d\lambda,$$

where the spectral shift function $\xi(\cdot; A_+, A_-)$ is fixed by (4.31).

Proof Since $|F'(\lambda)| \leq \text{const}(1+\lambda^2)^{-1}$, $\lambda \in \mathbb{R}$, Corollary 4.3.6 implies that

$$\int_{\mathbb{R}} F'(\lambda)\xi(\lambda; A_+, A_-)d\lambda = \lim_{n\to\infty} \int_{\mathbb{R}} F'(\lambda)\xi(\lambda; A_{+,n}, A_-)d\lambda.$$

Since $B_{+,n} = A_{+,n} - A_-$ is a trace-class operator (see (3.6)) and $F \in C_b^\infty(\mathbb{R})$, the Lipschitz-Krein trace formula implies that

$$\int_{\mathbb{R}} F'(\lambda)\xi(\lambda; A_{+,n}, A_-)d\lambda = \text{tr}\left(F(A_{+,n}) - F(A_-)\right).$$

Set $\psi = F \circ g^{-1}$. By the assumption on F and Theorem 2.2.10 the double operator integral $T_{\psi^{[1]}}^{g(A_{+,n}),g(A_-)}$ is bounded on $\mathcal{L}_1(\mathcal{H})$. Therefore,

$$\begin{aligned} F(A_{+,n}) - F(A_-) &= \psi(g(A_{+,n})) - \psi(g(A_-)) \\ &= T_{\psi^{[1]}}^{g(A_{+,n}),g(A_-)}\left(g(A_{+,n}) - g(A_-)\right) \end{aligned}$$

and similarly

$$F(A_+) - F(A_-) = T_{\psi^{[1]}}^{g(A_+),g(A_-)}\left(g(A_+) - g(A_-)\right)$$

In particular, we have that $F(A_+) - F(A_-) \in \mathcal{L}_1(\mathcal{H})$.

By the assumption on F, the function ψ satisfies the assumption of Proposition 2.4.4. Therefore, combining with Proposition 2.4.4 we obtain

$$\begin{aligned} \|\cdot\|_1 - \lim_{n\to\infty} \left(F(A_{+,n}) - F(A_-)\right) &= \|\cdot\|_1 - \lim_{n\to\infty} T_{\psi^{[1]}}^{g(A_{+,n}),g(A_-)}\left(g(A_{+,n}) - g(A_-)\right) \\ &= T_{\psi^{[1]}}^{g(A_+),g(A_-)}\left(g(A_+) - g(A_-)\right) \\ &= \psi(g(A_+)) - \psi(g(A_-)) = \left(F(A_+) - F(A_-)\right). \end{aligned}$$

Consequently we have

$$\lim_{n\to\infty} \operatorname{tr}\big(F(A_{+,n}) - F(A_-)\big) = \operatorname{tr}\big(F(A_+) - F(A_-)\big).$$

Finally,

$$\begin{aligned}\int_{\mathbb{R}} F'(\lambda)\xi(\lambda; A_+, A_-)d\lambda &= \lim_{n\to\infty} \int_{\mathbb{R}} F'(\lambda)\xi(\lambda; A_{+,n}, A_-)d\lambda \\ &= \lim_{n\to\infty} \operatorname{tr}\big(F(A_{+,n}) - F(A_-)\big) \\ &= \operatorname{tr}\big(F(A_+) - F(A_-)\big)\end{aligned}$$

as required. □

In addition to Proposition 7.2.4 we prove here a lemma where a useful integral decomposition for the difference $g(A_+) - g(A_-)$ is given. This integral decomposition is necessary in the proof that the assumptions of Proposition 7.2.4 are satisfied for the present example.

For brevity, we introduce the notation $R_{+,\lambda}(z)$, $R_{-,\lambda}(z)$ for appropriate resolvents of the operators A_+ and A_-, respectively, that is,

$$R_{+,\lambda} = \big(A_+ + i(1+\lambda)^{1/2}\big)^{-1},\ R_{-,\lambda} = \big(A_- + i(1+\lambda)^{1/2}\big)^{-1},\ \lambda > 0. \tag{7.14}$$

We also introduce

$$U_\lambda = (A_+ - A_-)R_{-,\lambda} = B_+R_{-,\lambda}. \tag{7.15}$$

Lemma 7.2.5 *Suppose that B_+ leaves the domain of A_- invariant then we have*

$$\begin{aligned}g(A_+) - g(A_-) &= -B_+(A_-^2 + 1)^{-3/2} \\ &+ \frac{1}{2\pi}\int_0^\infty \lambda^{-1/2}\big(R_{-,\lambda}[A_-, B_+]R_{-,\lambda}^2 + R_{-,\lambda}^*[A_-, B_+](R_{-,\lambda}^*)^2\big)d\lambda \\ &+ \frac{1}{\pi}\int_0^\infty \lambda^{-1/2}\mathrm{Re}\big(R_{+,\lambda}U_\lambda^2\big)d\lambda.\end{aligned}$$

Proof By [CGG⁺16, Lemma 3.1] we have

$$g(A_+) - g(A_-) = \frac{1}{\pi}\mathrm{Re}\Big(\int_0^\infty \lambda^{-1/2}[R_{+,\lambda} - R_{-,\lambda}]d\lambda\Big).$$

Using the resolvent identity twice one can write

$$\begin{aligned}R_{+,\lambda} - R_{-,\lambda} &= -R_{+,\lambda}B_+R_{-,\lambda} = -R_{-,\lambda}B_+R_{-,\lambda} + R_{+,\lambda}B_+R_{-,\lambda}B_+R_{-,\lambda} \\ &= -R_{-,\lambda}B_+R_{-,\lambda} + R_{+,\lambda}U_\lambda^2.\end{aligned}$$

Therefore, we have

$$R_{+,\lambda} - R_{-,\lambda} = -B_+ R_{-,\lambda}^2 - [R_{-,\lambda}, B_+]R_{-,\lambda} + R_{+,\lambda}U_\lambda^2.$$

Applying the identity $[C^{-1}, B] = -C^{-1}[C, B]C^{-1}$ to the second term we obtain

$$R_{+,\lambda} - R_{-,\lambda} = -B_+ R_{-,\lambda}^2 + R_{-,\lambda}[A_-, B_+]R_{-,\lambda}^2 + R_{+,\lambda}U_\lambda^2.$$

Similarly

$$\begin{aligned}
(R_{+,\lambda} - R_{-,\lambda})^* &= -R_{-,\lambda}^* + R_{+,\lambda}^* = -R_{-,\lambda}^* B_+ R_{+,\lambda}^* \\
&= -R_{-,\lambda}^* B_+ R_{-,\lambda}^* + R_{-,\lambda}^* B_+ R_{-,\lambda}^* B_+ R_{+,\lambda}^* \\
&= -R_{-,\lambda}^* B_+ R_{-,\lambda}^* + (R_{+,\lambda}U_\lambda^2)^* \\
&= -B_+(R_{-,\lambda}^*)^2 - [B_+, R_{-,\lambda}^*]R_{-,\lambda}^* + (R_{+,\lambda}U_\lambda^2)^* \\
&= -B_+(R_{-,\lambda}^*)^2 + R_{-,\lambda}^*[A_-, B_+](R_{-,\lambda}^*)^2 + (R_{+,\lambda}U_\lambda^2)^*.
\end{aligned}$$

Therefore, we have

$$\begin{aligned}
g(A_+) - g(A_-) &= \frac{1}{2\pi}\int_0^\infty \lambda^{-1/2} B_+\big((R_{-,\lambda}^*)^2 + (R_{-,\lambda})^2\big)d\lambda \\
&+ \frac{1}{2\pi}\int_0^\infty \lambda^{-1/2}\big(R_{-,\lambda}[A_-, B_+]R_{-,\lambda}^2 + R_{-,\lambda}^*[A_-, B_+](R_{-,\lambda}^*)^2\big)d\lambda \\
&+ \frac{1}{2\pi}\int_0^\infty \lambda^{-1/2}\big(R_{+,\lambda}U_\lambda^2 + (R_{+,\lambda}U_\lambda^2)^*\big)d\lambda \\
&= \frac{1}{\pi}B_+\mathrm{Re}\Big(\int_0^\infty \lambda^{-1/2}(R_{-,\lambda})^2 d\lambda\Big) \\
&+ \frac{1}{2\pi}\int_0^\infty \lambda^{-1/2}\big(R_{-,\lambda}[A_-, B_+]R_{-,\lambda}^2 + R_{-,\lambda}^*[A_-, B_+](R_{-,\lambda}^*)^2\big)d\lambda \\
&+ \frac{1}{\pi}\int_0^\infty \lambda^{-1/2}\mathrm{Re}\Big(R_{+,\lambda}U_\lambda^2\Big)d\lambda.
\end{aligned} \tag{7.16}$$

Now,

$$\begin{aligned}
\int_0^\infty \frac{d\lambda}{\lambda^{1/2}}\mathrm{Re}(R_{0,\lambda}^2) &= \int_0^\infty \frac{d\lambda}{\lambda^{1/2}}(A_-^2 - (1+\lambda)(1+\lambda+A_-^2)^{-2}) \\
&= (A_-^2 - 1)\int_0^\infty \lambda^{-1/2}(1+\lambda+A_-^2)^{-2}\,d\lambda - \int_0^\infty \lambda^{1/2}(1+\lambda+A_-^2)^{-2}\,d\lambda \\
&= \frac{\pi}{2}(A_-^2 - 1)(1+A_-^2)^{-3/2} - \frac{\pi}{2}(1+A_-^2)^{-1/2} = -\pi(1+A_-^2)^{-3/2},
\end{aligned}$$

and hence the first term on the right hand side of (7.16) can be written as

$$\frac{1}{\pi}B_+\mathrm{Re}\Big(\int_0^\infty \lambda^{-1/2}(R_{-,\lambda})^2 d\lambda\Big) = -B_+(A_-^2+1)^{-3/2},$$

which suffices to complete the proof. □

Proposition 7.2.6 *Let $\varphi \in C^1[0,2\pi]$ with $\varphi(0)=\varphi(2\pi)$. Then $g(D_\alpha+\varphi)-g(D_\alpha)$ is trace-class. In addition, if $P_n = \chi_{[-n,n]}(D_\alpha)$ then*

$$\lim_{n\to\infty}\Big\|[g(D_\alpha+P_n\varphi P_n)-g(D_\alpha)]-[g(D_\alpha+\varphi)-g(D_\alpha)]\Big\|_1 = 0.$$

Proof Using the decomposition obtained in Lemma 7.2.5 we can write

$$\begin{aligned} g(D_\alpha+\varphi)-g(D_\alpha) &= -\varphi(D_\alpha^2+1)^{-3/2} \\ &+\frac{1}{2\pi}\int_0^\infty \lambda^{-1/2}\big(R_{-,\lambda}[D_\alpha,\varphi]R_{-,\lambda}^2 + R_{-,\lambda}^*[D_\alpha,\varphi](R_{-,\lambda}^*)^2\big)d\lambda \\ &+\frac{1}{\pi}\int_0^\infty \lambda^{-1/2}\mathrm{Re}\big(R_{+,\lambda}U_\lambda^2\big)d\lambda. \end{aligned} \tag{7.17}$$

where, as before,

$$R_{+,\lambda} = (D_\alpha+\varphi+i(\lambda+1))^{-1/2},\ R_{-,\lambda} = (D_\alpha+i(\lambda+1))^{-1/2},\ U_\lambda = \varphi R_{-,\lambda},\quad \lambda>0.$$

We note that both integrals on the right-hand side converge in the uniform norm.

The first term on the right hand side of (7.17) is a trace-class operator, since

$$\Big\|\varphi(D_\alpha^2+1)^{-3/2}\Big\|_1 \le \|\varphi\|\Big\|(D_\alpha^2+1)^{-3/2}\Big\|_1 = \|\varphi\|\sum_{n\in\mathbb{Z}}\frac{1}{((n-\alpha)^2+1)^{3/2}} < \infty.$$

For the second term in (7.17) we note that since the function φ is differentiable and $\varphi(0)=\varphi(2\pi)$ we have that the commutator $[D_\alpha,\varphi]$ extends to a bounded operator on $L_2[0,2\pi]$ (which we still denote by $[D_\alpha,\varphi]$). Hence, we can estimate

$$\begin{aligned} \Big\|\int_0^\infty \frac{d\lambda}{\lambda^{1/2}}R_{-,\lambda}[D_\alpha,\varphi]R_{-,\lambda}^2\Big\|_1 &\le \int_0^\infty \frac{d\lambda}{\lambda^{1/2}}\|R_{-,\lambda}[D_\alpha,\varphi]\|_3\|R_{-,\lambda}^2\|_{3/2} \\ &\le \int_0^\infty \frac{d\lambda}{\lambda^{1/2}}\|[D_\alpha,\varphi]\|\Big\|(D_\alpha+i(\lambda+1)^{-1/2})^{-1}\Big\|_3^3 \\ &\le \mathrm{const}\int_0^\infty \frac{d\lambda}{\lambda^{1/2}}\sum_{n\in\mathbb{Z}}\frac{1}{((n-\alpha)^2+(1+\lambda))^{3/2}} \\ &\le \mathrm{const}\int_0^\infty \frac{d\lambda}{\lambda^{1/2}}(1+\lambda)^{-1} < \infty, \end{aligned}$$

and therefore,

$$\int_0^\infty \frac{d\lambda}{\lambda^{1/2}} R_{-,\lambda}[D_\alpha, \varphi]R_{-,\lambda}^2 \in \mathcal{L}_1(L_2[0, 2\pi]).$$

A similar argument shows that

$$\int_0^\infty \frac{d\lambda}{\lambda^{1/2}} R_{-,\lambda}^*[D_\alpha, \varphi](R_{-,\lambda}^2)^* \in \mathcal{L}_1(L_2[0, 2\pi]).$$

That is, the second term in (7.17) is also a trace-class operator.

For the third term in (7.17) we have

$$\Big\| \int_0^\infty \frac{d\lambda}{\lambda^{1/2}} R_{+,\lambda} U_\lambda^2 \Big\|_1 \le \int_0^\infty \frac{d\lambda}{\lambda^{1/2}} \|R_{+,\lambda}\| \|U_\lambda\|_2^2$$

and since $\|R_{+,\lambda}\| \le \mathrm{const}(1+\lambda)^{-1/2}$ and

$$\begin{aligned} \|U_\lambda\|_2^2 &= \Big\|\varphi(D_\alpha + i(\lambda+1)^{-1/2})^{-1}\Big\|_2^2 \le \|\varphi\|_\infty^2 \Big\|(D_\alpha + i(\lambda+1)^{-1/2})^{-1}\Big\|_2^2 \\ &\le \mathrm{const} \sum_{n\in\mathbb{Z}} \frac{1}{(n-\alpha)^2 + 1 + \lambda} \le \mathrm{const}(1+\lambda)^{-1/2}. \end{aligned}$$

we infer that

$$\Big\| \int_0^\infty \frac{d\lambda}{\lambda^{1/2}} R_{+,\lambda} U_\lambda^2 \Big\|_1 \le \mathrm{const} \int_0^\infty \frac{d\lambda}{\lambda^{1/2}} (1+\lambda)^{-1} < \infty.$$

Thus all terms in (7.17) are trace-class operators.

The proof of convergence follows an argument similar to that of [CGG[+]16, Theorem 3.5], and therefore is omitted. □

Remark 7.2.7 We note that by equality (7.17) for sufficiently small $t \in \mathbb{R}$ we have $g(D_\alpha + t\varphi) - g(D_\alpha) = tV + o^{\mathcal{L}_1}(t)$, where

$$V = \varphi(D_\alpha^2 + 1)^{-3/2} - \int_0^\infty \frac{d\lambda}{\lambda^{1/2}} \big(R_{-,\lambda}[D_\alpha, \varphi]R_{-,\lambda}^2 + R_{-,\lambda}^*[D_\alpha, \varphi](R_{-,\lambda}^2)^*\big).$$

The assumption that $\varphi(0) = \varphi(2\pi)$ is required for Proposition 7.2.6, since its proof uses the commutator $[D_\alpha, \varphi]$, which is well-defined only if $\varphi(0) = \varphi(2\pi)$. ◇

As the assumptions of Proposition 7.2.4 are satisfied it now follows that

$$\mathrm{tr}\big(F(D_\alpha + \varphi) - F(D_\alpha)\big) = \int_{\mathbb{R}} F'(\lambda)\xi(\lambda; D_\alpha + \varphi, D_\alpha)d\lambda \tag{7.18}$$

provided that F is such that $F' \in S(\mathbb{R})$. To compute the spectral shift function $\xi(\cdot, D_\alpha + \varphi, D_\alpha)$ we shall compute the left hand of the equality above for any F with $F' \in S(\mathbb{R})$.

Let $\varphi \in C^1[0, 2\pi]$. Recall that

$$c := \frac{1}{2\pi}\int_0^{2\pi} \varphi(s)ds, \quad \varphi_0 := \varphi - c \quad \text{and} \int_0^{2\pi} \varphi_0(s)ds = 0. \tag{7.19}$$

We write

$$\mathrm{tr}(F(D_\alpha+\varphi)-F(D_\alpha)) = \mathrm{tr}(F(D_\alpha+\varphi_0+c)-F(D_\alpha+\varphi_0))+\mathrm{tr}(F(D_\alpha+\varphi_0)-F(D_\alpha)).$$

Since $\int_0^{2\pi} \varphi_0(s)ds = 0$, Lemma 7.2.1 implies that the operator $D_\alpha + \varphi_0$ is unitarily equivalent to the operator D_α. Therefore, we have

$$\mathrm{tr}(F(D_\alpha + \varphi) - F(D_\alpha)) \qquad = \mathrm{tr}(F(D_\alpha + c) - F(D_\alpha))+\mathrm{tr}(F(D_\alpha + \varphi_0) - F(D_\alpha)). \tag{7.20}$$

We claim that $\mathrm{tr}(F(D_\alpha + \varphi_0) - F(D_\alpha)) = 0$. Prior to proving this equality, we establish the following auxiliary result.

Lemma 7.2.8 *If $\Psi \in S(\mathbb{R})$, then $\Psi(D_\alpha) \in \mathcal{L}_1(L_2[0, 2\pi])$. Moreover, if $\varphi_0 \in C^1[0, 2\pi]$ is such that $\varphi_0(0) = \varphi_0(2\pi)$ and $\int_0^{2\pi} \varphi_0(s)ds = 0$, then* $\mathrm{tr}(\varphi_0\Psi(D_\alpha)) = 0$.

Proof Since $\Psi \in S(\mathbb{R})$, there exists $C \geq 0$, such that $|\Psi(s)| \leq C(1+s^2)^{-1}$, $s \in \mathbb{R}$. Therefore,

$$\|\Psi(D_\alpha)\|_1 = \sum_{n\in\mathbb{Z}} |\Psi(\lambda_{\alpha,n})| \leq C\sum_{n\in\mathbb{Z}} \frac{1}{1 + (n - \alpha)^2} < \infty,$$

that is $\Psi(D_\alpha) \in \mathcal{L}_1$.

By the unitary equivalence given in (7.12) we have

$$\mathrm{tr}(\varphi_0\Psi(D_\alpha)) = \mathrm{tr}\left(U_\alpha\varphi_0\Psi\big(D_0 - \alpha\big)U_\alpha^*\right) = \mathrm{tr}\left(\varphi_0\Psi\big(D_0 - \alpha\big)\right). \tag{7.21}$$

Denoting $\lambda_n := \lambda_{0,n}$ and $e_n := e_{0,n}$ (here, $\lambda_{0,n}$ and $e_{0,n}$ are given by (7.11) with $\alpha = 0$) for an arbitrary $\xi \in L_2[0, 2\pi]$ we can write

$$\begin{aligned}\varphi_0\Psi\big(D_0 - \alpha\big)\xi &= \sum_{n_1\in\mathbb{Z}} \hat{\varphi}_0(n_1)e_{n_1} \cdot \sum_{n_2\in\mathbb{Z}} \Psi\big(\lambda_n - \alpha\big)\hat{\xi}(n_2)e_{n_2} \\ &= \sum_{n\in\mathbb{Z}} e_n \sum_{n_1+n_2=n} \Psi\big(\lambda_n - \alpha\big)\hat{\xi}(n_2)\hat{\varphi}_0(n_1),\end{aligned}$$

where $\hat{\eta}(n)$ denotes the n-th Fourier coefficient of a function $\eta \in L_2[0, 2\pi]$, $\eta(0) = \eta(2\pi)$ and the series converge in $L_2[0, 2\pi]$ norm. That is

$$\left(\widehat{\varphi_0 \Psi(D_0 - \alpha)\xi}\right)(n) = \sum_{n_2 \in \mathbb{Z}} \hat{\xi}(n_2)\Psi\big(\lambda_n - \alpha\big)\hat{\varphi}_0(n - n_2).$$

Hence, the matrix elements $K(n, n_2)$ of the operator $\varphi_0\Psi(D - \alpha)$ are given by $K(n, n_2) = \Psi(\lambda_n - \alpha)\hat{\varphi}_0(n - n_2)$, and therefore,

$$\begin{aligned}\operatorname{tr}\Big(\varphi_0\Psi\big(D_0 - \alpha\big)\Big) &= \sum_{n \in \mathbb{Z}} K(n, n) = \sum_{n \in \mathbb{Z}} \Psi\big(\lambda_n - \alpha\big)\hat{\varphi}_0(0) \\ &= \sum_{n \in \mathbb{Z}} \Psi\big(\lambda_n - \alpha\big) \int_0^{2\pi} \varphi_0(s)ds \overset{(7.19)}{=} 0,\end{aligned}$$

as required. □

Lemma 7.2.9 *Let F be such that $F' \in S(\mathbb{R})$ and let $\varphi_0 \in C^1[0, 2\pi]$, $\varphi_0(0) = \varphi_0(2\pi)$ with $\int_0^{2\pi} \varphi_0(s)ds = 0$. Then* $\operatorname{tr}(F(D_\alpha + \varphi_0) - F(D_\alpha)) = 0$.

Proof We set

$$H(t) = F(D_\alpha + tM_{\varphi_0}) - F(D_\alpha), \quad t \in \mathbb{R}.$$

Since $t\varphi_0 \in C^1[0, 2\pi]$, $t\varphi_0(0) = t\varphi_0(2\pi)$ for all $t \in \mathbb{R}$, Proposition 7.2.6 implies that $H(t) \in \mathcal{L}_1(L_2[0, 2\pi])$ for all $t \in \mathbb{R}$. We claim that the $\mathcal{L}_1$-valued function $H(t)$ is differentiable in $\mathcal{L}_1$-norm.

Let $t, t_0 \in \mathbb{R}$. As $\int_0^{2\pi} t_0\varphi_0(s)ds = 0$ for all $t_0 \in \mathbb{R}$ the operator $D_\alpha + t_0M_{\varphi_0}$ is unitarily equivalent to the operator D_α via the operator ψ_{t_0}, where $\psi_{t_0}(\nu) = \exp(-i\int_0^s t_0\varphi_0(s)ds)$. In addition, since ψ_{t_0} commutes with φ_0, we also have that

$$D_\alpha + t_0\varphi_0 + (t - t_0)\varphi_0 = \psi_{t_0}(D_\alpha + (t - t_0)\varphi_0)\bar{\psi}_{t_0}.$$

Therefore,

$$\begin{aligned}H(t) - H(t_0) &= F(D_\alpha + tM_{\varphi_0}) - F(D_\alpha + t_0M_{\varphi_0}) \\ &= F\big(D_\alpha + t_0M_{\varphi_0} + (t - t_0)\varphi_0\big) - F(D_\alpha + t_0M_{\varphi_0}) \\ &= \psi_{t_0}\Big(F(D_\alpha + (t - t_0)\varphi_0) - F(D_\alpha)\Big)\bar{\psi}_{t_0} = \psi_{t_0}H(t - t_0)\bar{\psi}_{t_0}.\end{aligned} \tag{7.22}$$

Thus, it is sufficient to prove differentiability of $H(t)$ at $t_0 = 0$ only.

In the following we choose t sufficiently small. Setting $G = F \circ g^{-1}$, then G is an infinitely differentiable bounded function on $(-1, 1)$. Now, by Remark 7.2.7 we have

$$\begin{aligned} H(t) &= G\big(g(D_\alpha + tM_{\varphi 0})\big) - G\big(g(D_\alpha)\big) \\ &= T_{G^{[1]}}^{g(D_\alpha + tM_{\varphi 0}), g(D_\alpha)}\Big(g(D_\alpha + tM_{\varphi 0}) - g(D_\alpha)\Big) \\ &= T_{G^{[1]}}^{g(D_\alpha + tM_{\varphi 0}), g(D_\alpha)}\Big(tV + o^{\mathcal{L}_1}(t)\Big) \\ &= tT_{G^{[1]}}^{g(D_\alpha + tM_{\varphi 0}), g(D_\alpha)}(V) + T_{G^{[1]}}^{g(D_\alpha + tM_{\varphi 0}), g(D_\alpha)}\Big(o^{\mathcal{L}_1}(t)\Big). \end{aligned}$$

Hence

$$\frac{d}{dt}H(t)\Big|_{t=0} = \|\cdot\|_1\text{-}\lim_{t\to 0} T_{G^{[1]}}^{g(D_\alpha + tM_{\varphi 0}), g(D_\alpha)}(V) + \|\cdot\|_1\text{-}\lim_{t\to 0} \frac{1}{t} T_{G^{[1]}}^{g(D_\alpha + tM_{\varphi 0}), g(D_\alpha)}\Big(o^{\mathcal{L}_1}(t)\Big)$$

From Proposition 2.4.4 it follows that

$$\frac{d}{dt}H(t)\Big|_{t=0} = T_{G^{[1]}}^{g(D_\alpha), g(D_\alpha)}(V) \in \mathcal{L}_1(\mathcal{H}),$$

that is, the function $H(t)$ is differentiable in $\mathcal{L}_1$-norm. In particular, by (7.22) it follows that

$$\frac{d}{dt}H(t)\Big|_{t=t_0} = \psi_{t_0} T_{G^{[1]}}^{g(D_\alpha), g(D_\alpha)}(V)\bar{\psi}_{t_0}. \tag{7.23}$$

Next, by Propositions 2.4.6 and 2.2.5 we have

$$\begin{aligned} &\operatorname{tr}(F(D_\alpha + M_h) - F(D_\alpha)) = \operatorname{tr}(H(1)) \\ &= \operatorname{tr}\Big(\int_0^1 \frac{d}{dt}H(t)dt\Big) = \int_0^1 \operatorname{tr}(T_{G^{[1]}}^{g(D_\alpha), g(D_\alpha)}(V))dt \\ &= \operatorname{tr}(T_{G^{[1]}}^{g(D_\alpha), g(D_\alpha)}(V)) \overset{(2.6)}{=} \operatorname{tr}(T_{G^{[1]}}^{g(D_\alpha), g(D_\alpha)}(1)\cdot V) \\ &= \operatorname{tr}(G'(g(D_\alpha))V). \end{aligned} \tag{7.24}$$

It remains now to compute the trace $\operatorname{tr}(G'(g(D_\alpha))V)$. Recall that

$$V = \varphi(D_\alpha^2 + 1)^{-3/2} - \int_0^\infty \frac{d\lambda}{\lambda^{1/2}}\big(R_{-,\lambda}[D_\alpha, \varphi]R_{-,\lambda}^2 + R_{-,\lambda}^*[D_\alpha, \varphi](R_{-,\lambda}^2)^*\big),$$

where every separate term is a trace-class operator. Since the function $G' \circ g$ is a Schwartz function, it follows from Lemma 7.2.8 that $G'(g(D_\alpha)) \in \mathcal{L}_1(L_2[0, 2\pi])$.

Therefore, again using Lemma 7.2.8 we have

$$\begin{aligned} \operatorname{tr}(G'(g(D_\alpha))V) &= \operatorname{tr}\left(G'(g(D_\alpha))\varphi_0(D_\alpha^2+1)^{-3/2}\right) \\ &\quad - \int_0^\infty \frac{d\lambda}{\lambda^{1/2}} \operatorname{tr}\left(G'(g(D_\alpha))R_{-,\lambda}[D_\alpha,\varphi_0]R_{-,\lambda}^2\right) \\ &\quad - \int_0^\infty \frac{d\lambda}{\lambda^{1/2}} \operatorname{tr}\left(G'(g(D_\alpha))R_{-,\lambda}^*[D_\alpha,\varphi_0](R_{-,\lambda}^2)^*\right). \end{aligned} \tag{7.25}$$

Since $G'(g(D_\alpha)) \in \mathcal{L}_1(L_2[0,2\pi])$, we have that

$$\operatorname{tr}\left(G'(g(D_\alpha))\varphi_0(D_\alpha^2+1)^{-3/2}\right) = \operatorname{tr}\left(\varphi_0 G'(g(D_\alpha))(D_\alpha^2+1)^{-3/2}\right) = 0,$$

where the last equality follows from Lemma 7.2.8. Using a similar argument (while opening the commutator) one can show that the second and third terms in (7.25) are also 0. Hence, we infer from (7.24) that $\operatorname{tr}(F(D_\alpha+\varphi_0) - F(D_\alpha)) = 0$. □

Now, we are ready to compute explicitly the spectral shift function $\xi(\cdot; D_\alpha + \varphi, D_\alpha)$.

Theorem 7.2.10 *Let* $D_\alpha = \frac{d}{idx}$ *with* α*-twisted periodic boundary conditions on* $[0,2\pi]$, $\alpha \in [0,2\pi)$, *and let* $\varphi \in C^1[0,2\pi]$, $\varphi(0) = \varphi(2\pi)$, $c = \frac{1}{2\pi}\int_0^{2\pi} \varphi(s)ds$. *Then*

$$\xi(\cdot; D_\alpha + \varphi, D_\alpha) = \lfloor c \rfloor + \operatorname{sgn}(c) \sum_{n\in\mathbb{Z}} \chi_{(\lambda_{\alpha,n},\lambda_{\alpha,n}+\{c\})} \text{ a.e..}$$

Proof Let F be an arbitrary function with $F' \in S(\mathbb{R})$. By (7.18) we have

$$\operatorname{tr}(F(D_\alpha+\varphi) - F(D_\alpha)) = \int_{\mathbb{R}} F'(s)\xi(s; D_\alpha+\varphi, D_\alpha)ds.$$

By equality (7.20) and Lemma 7.2.9 we have that

$$\operatorname{tr}(F(D_\alpha+\varphi) - F(D_\alpha)) = \operatorname{tr}(F(D_\alpha+c) - F(D_\alpha)),$$

and therefore, it is sufficient to compute the trace $\operatorname{tr}(F(D_\alpha+c) - F(D_\alpha))$. We have

$$\begin{aligned} \operatorname{tr}(F(D_\alpha+c) - F(D_\alpha)) &= \sum_{n\in\mathbb{Z}} \left(F(\lambda_{\alpha,n}+c) - F(\lambda_{\alpha,n})\right) \\ &= \sum_{n\in\mathbb{Z}} \int_{\lambda_{\alpha,n}}^{\lambda_{\alpha,n}+c} F'(s)ds. \end{aligned}$$

Therefore, for an arbitrary F such that $F' \in S(\mathbb{R})$ we have

$$\int_{\mathbb{R}} F'(s)\xi(s; D_\alpha + \varphi, D_\alpha)ds = \sum_{n\in\mathbb{Z}} \int_{\lambda_{\alpha,n}}^{\lambda_{\alpha,n}+c} F'(s)ds \tag{7.26}$$

$$= \mathrm{sgn}(c) \int_{\mathbb{R}} \sum_{n\in\mathbb{Z}} \chi_{(\lambda_{\alpha,n},\lambda_{\alpha,n}+c)}(s)F'(s)ds.$$

Then (7.26) suffices to show that

$$\xi(\cdot; D_\alpha + \varphi, D_\alpha) = \mathrm{sgn}(c) \sum_{n\in\mathbb{Z}} \chi_{(\lambda_{\alpha,n},\lambda_{\alpha,n}+c)} \quad \text{a.e.},$$

where $\chi_{(\lambda_{\alpha,n},\lambda_{\alpha,n}+c)}$ stands for the characteristic function of the set $(\lambda_{\alpha,n} + c, \lambda_{\alpha,n})$ if $c < 0$.

Therefore,

$$\begin{aligned}
\xi(\cdot; D_\alpha + \varphi, D_\alpha) &= \mathrm{sgn}(c) \sum_{n\in\mathbb{Z}} \chi_{(\lambda_{\alpha,n},\lambda_{\alpha,n}+c)} \\
&= \mathrm{sgn}(c) \sum_{n\in\mathbb{Z}} \chi_{(\lambda_{\alpha,n},\lambda_{\alpha,n+\lfloor c\rfloor})} \\
&\quad + \mathrm{sgn}(c) \sum_{n\in\mathbb{Z}} \chi_{(\lambda_{\alpha,n+\lfloor c\rfloor},\lambda_{\alpha,n+\lfloor c\rfloor}+\{c\})} \\
&= \mathrm{sgn}(c)\big|\lfloor c\rfloor\big| + \mathrm{sgn}(c) \sum_{n\in\mathbb{Z}} \chi_{(\lambda_{\alpha,n},\lambda_{\alpha,n}+\{c\})} \\
&= \lfloor c\rfloor + \mathrm{sgn}(c) \sum_{n\in\mathbb{Z}} \chi_{(\lambda_{\alpha,n},\lambda_{\alpha,n}+\{c\})}.
\end{aligned}$$

□

7.2.3 The Index of the Operator D_A

Having computed explicitly the spectral shift function $\xi(\cdot; D_\alpha + \varphi, D_\alpha)$ we now compute the Witten index of the corresponding operator $\boldsymbol{D_A} = \frac{d}{dt} \otimes 1 + 1 \otimes D_\alpha + \theta \otimes \varphi$.

Theorem 7.2.11 *Let* $\varphi \in C^1[0, 2\pi]$, $\varphi(0) = \varphi(2\pi)$, $c = \frac{1}{2\pi}\int_0^{2\pi} \varphi(s)ds$. *Let* θ *satisfy* (3.33) *and let* $D_\alpha = \frac{d}{idx}$ *on* $L_2[0, 2\pi]$ *with* α*-twisted periodic boundary* *conditions,* $\alpha \in [0, 2\pi)$*. Then the Witten indices* $W_s(\boldsymbol{D}_{\boldsymbol{A}})$*,* $W_{k,r}(\boldsymbol{D}_{\boldsymbol{A}})$*,* $k \geq 1$*, exist and we have*

(i) If $\alpha = 0$*, then the operator* $\boldsymbol{D}_{\boldsymbol{A}}$ *is not Fredholm for any* φ *and*

$$W_s(\boldsymbol{D}_{\boldsymbol{A}}) = W_{k,r}(\boldsymbol{D}_{\boldsymbol{A}}) = \begin{cases} \lfloor c \rfloor, & \text{if } c \in \mathbb{Z} \\ \lfloor c \rfloor + \frac{1}{2}\operatorname{sgn}(c), & \text{otherwise.} \end{cases}$$

(ii) If $\alpha \neq 0$*, then the operator* $\boldsymbol{D}_{\boldsymbol{A}}$ *is Fredholm if and only if* $c \notin \alpha + \mathbb{Z}$*. In this case*

$$W_s(\boldsymbol{D}_{\boldsymbol{A}}) = W_{k,r}(\boldsymbol{D}_{\boldsymbol{A}}) = \operatorname{index}(\boldsymbol{D}_{\boldsymbol{A}}) = \begin{cases} \lfloor c \rfloor + \operatorname{sgn}(c), & \{c\} > \alpha \\ \lfloor c \rfloor, & \{c\} < \alpha. \end{cases}$$

If $c \in \alpha + \mathbb{Z}$*, then the operator* $\boldsymbol{D}_{\boldsymbol{A}}$ *is not Fredholm and* $W_s(\boldsymbol{D}_{\boldsymbol{A}}) = \lfloor c \rfloor + \frac{1}{2}\operatorname{sgn}(c)$*.*

Proof If $\alpha = 0$, then $0 \in \sigma(D)$, hence, by Theorem 3.2.3 the operator $\boldsymbol{D}_{\boldsymbol{A}}$ is not Fredholm. If $\alpha \neq 0$, then $0 \notin \sigma(D_\alpha)$, and therefore $\boldsymbol{D}_{\boldsymbol{A}}$ is Fredholm if and only if $0 \notin \sigma(D_\alpha + \varphi)$. By Remark 7.2.2 we have that

$$\sigma(D_\alpha + \varphi) = \Big\{n - \alpha + c, n \in \mathbb{Z}\Big\}.$$

Hence $0 \in \sigma(D_\alpha + \varphi)$ if and only if $c \in \alpha + \mathbb{Z}$. Thus, if $\alpha \neq 0$, the operator $\boldsymbol{D}_{\boldsymbol{A}}$ is Fredholm if and only if $c \notin \alpha + \mathbb{Z}$.

Now, we turn to computing the index of the operator $\boldsymbol{D}_{\boldsymbol{A}}$. It follows from Theorem 7.2.10 that the spectral shift function $\xi(\,\cdot\,; D_\alpha + \varphi, D_\alpha)$ is piecewise constant, and therefore, 0 is a right and a left Lebesgue point of $\xi(\,\cdot\,; D_\alpha + \varphi, D_\alpha)$. By Theorem 6.4.11 the Witten indices $W_s(\boldsymbol{D}_{\boldsymbol{A}})$, $W_{k,r}(\boldsymbol{D}_{\boldsymbol{A}})$, $k \geq 1$, exist and equal

$$\begin{aligned} W_s(\boldsymbol{D}_{\boldsymbol{A}}) = W_{k,r}(\boldsymbol{D}_{\boldsymbol{A}}) &= [\xi_L(0_+; D_\alpha + \varphi, D_\alpha) + \xi_L(0_-; D_\alpha + \varphi, D_\alpha)]/2 \\ &= [\xi(0_+; D_\alpha + \varphi, D_\alpha) + \xi(0_-; D_\alpha + \varphi, D_\alpha)]/2. \end{aligned} \tag{7.27}$$

By Theorem 7.2.10 we have

$$\xi(\cdot; D_\alpha + \varphi, D_\alpha) = \lfloor c \rfloor + \operatorname{sgn}(c) \sum_{n \in \mathbb{Z}} \chi_{(\lambda_{\alpha,n}, \lambda_{\alpha,n} + \{c\})},$$

and therefore

$$\xi(0_\pm; D_\alpha + \varphi, D_\alpha) = \lfloor c \rfloor + \operatorname{sgn}(c) \chi_{(\lambda_{\alpha,0}, \lambda_{\alpha,0} + \{c\})}(0_\pm). \tag{7.28}$$

Now we consider the cases $\alpha = 0$ and $\alpha \neq 0$ separately. Assume first that $\alpha = 0$, then by (7.28) we have

$$\xi(0_\pm; D_\alpha + \varphi, D_\alpha) = \lfloor c \rfloor + \mathrm{sgn}(c)\chi_{(0,\{c\})}(0\pm).$$

Thus,

$$\xi(0_+; D_\alpha + \varphi, D_\alpha) = \begin{cases} \lfloor c \rfloor + \mathrm{sgn}(c), & \{c\} \neq 0, \\ \lfloor c \rfloor, & \text{otherwise.} \end{cases}$$

and

$$\xi(0_-; D_\alpha + \varphi, D_\alpha) = \lfloor c \rfloor.$$

Hence when $\alpha = 0$ we infer from (7.27) that

$$W_s(\boldsymbol{D_A}) = W_{k,r}(\boldsymbol{D_A}) = \begin{cases} \lfloor c \rfloor, & \text{if } c \in \mathbb{Z} \\ \lfloor c \rfloor + \frac{1}{2}\,\mathrm{sgn}(c), & \text{otherwise.} \end{cases}$$

Now, let $\alpha \neq 0$. Then by (7.28) we have

$$\xi(0_\pm; D_\alpha + \varphi, D_\alpha) = \lfloor c \rfloor + \mathrm{sgn}(c)\chi_{(-\alpha,-\alpha+\{c\})}(0\pm).$$

Thus,

$$\xi(0_+; D_\alpha + \varphi, D_\alpha) = \begin{cases} \lfloor c \rfloor, & \{c\} \leq \alpha \\ \lfloor c \rfloor + \mathrm{sgn}(c), & \{c\} > \alpha, \end{cases}$$

and

$$\xi(0_-; D_\alpha + \varphi, D_\alpha) = \begin{cases} \lfloor c \rfloor, & \{c\} < \alpha \\ \lfloor c \rfloor + \mathrm{sgn}(c), & \{c\} \geq \alpha. \end{cases}$$

Combining these two equalities with equality (7.27) we obtain the following precise value of the Witten index of the operator $\boldsymbol{D_A}$

$$W_s(\boldsymbol{D_A}) = W_{k,r}(\boldsymbol{D_A}) = \begin{cases} \lfloor c \rfloor, & \{c\} < \alpha \\ \lfloor c \rfloor + \frac{1}{2}\,\mathrm{sgn}(c), & \{c\} = \alpha \\ \lfloor c \rfloor + \mathrm{sgn}(c), & \{c\} > \alpha. \end{cases}$$

□

Remark 7.2.12 It follows from Theorem 7.2.11, that if $\alpha \neq 0$ and $c \notin \alpha + \mathbb{Z}$, then we are in the Fredholm situation (i.e. the operators $D_\alpha + \varphi$, D_α and $\boldsymbol{D}_{\boldsymbol{A}}$ are Fredholm) with discrete spectra as in [RS95].

It is worth noting that in this special compact case the Witten index $W_s(\boldsymbol{D}_{\boldsymbol{A}})$ can take only half-integer values, while for the locally compact case (i.e. the operator $D = \frac{d}{idx}$ acts on $L_2(\mathbb{R})$) the Witten index $W_s(\boldsymbol{D}_{\boldsymbol{A}})$ could be any real number (see [CGL+16b] and [CGG+16]). ◇

7.2.4 Spectral Flow Along the path $\{D_\alpha + \theta(t)\varphi\}_{t\in\mathbb{R}}$

Now, we compute the spectral flow along the path $\{D_\alpha + \theta(t)\varphi\}_{t\in\mathbb{R}}$.

By Theorem 6.4.12 we have

$$\begin{aligned} \mathrm{sf}(D_\alpha, D_\alpha + \varphi) &= \frac{1}{2}\big(\xi(0_+; D_\alpha + \varphi, D_\alpha) + \xi(0_-; D_\alpha + \varphi, D_\alpha)\big) \\ &\quad + \frac{1}{2}\big(\dim(\ker(D_\alpha + \varphi)) - \dim(\ker(D_\alpha))\big). \end{aligned}$$

We again consider the cases $\alpha = 0$ and $\alpha \neq 0$ separately. First let $\alpha = 0$. It is clear that $\dim(\ker(D_\alpha)) = 1$. By Remark 7.2.2 we have

$$\dim(\ker(D_\alpha + \varphi)) = \dim(\ker(D_\alpha + c)) = \begin{cases} 1, & \text{if } c \in \mathbb{Z} \\ 0, & \text{otherwise.} \end{cases}$$

Thus, combining these equalities with Theorem 7.2.11 we obtain

$$\mathrm{sf}(D_\alpha, D_\alpha + \varphi) = \begin{cases} \lfloor c \rfloor, & c \in \mathbb{Z} \\ \lfloor c \rfloor + \frac{1}{2}\,\mathrm{sgn}(c) - \frac{1}{2} & \text{otherwise.} \end{cases}$$

Now, let $\alpha \neq 0$. If $c \notin \alpha + \mathbb{Z}$, then $0 \in \rho(D_\alpha) \cap \rho(D_\alpha + \varphi)$, and therefore, $\dim(\ker(D_\alpha + \varphi)) = \dim(\ker(D_\alpha)) = 0$. In addition, by Theorem 7.2.11 we are in the Fredholm situation and have the equality

$$W_r(\boldsymbol{D}_{\boldsymbol{A}}) = \mathrm{index}(\boldsymbol{D}_{\boldsymbol{A}}) = \mathrm{sf}(D_\alpha, D_\alpha + \varphi),$$

which is consistent with the result of [RS95] and [GLM+11, Theorem 9.13]. If $c = \alpha + \mathbb{Z}$, then $\dim(\ker(D_\alpha + \varphi)) = 1$, and by Theorem 7.2.11 we obtain

$$\mathrm{sf}(D_\alpha, D_\alpha + \varphi) = \lfloor c \rfloor + \frac{1}{2}\,\mathrm{sgn}(c) + 1/2.$$

□

Thus, we have the following

Theorem 7.2.13 *Let $\varphi \in C^1[0, 2\pi]$, $\varphi(0) = \varphi(2\pi)$, let θ satisfy* (3.33) *and let $D_\alpha = \frac{d}{idx}$ on $L_2[0, 2\pi]$ with α-twisted periodic boundary conditions, $\alpha \in [0, 2\pi)$. With $c = \frac{1}{2\pi}\int_0^{2\pi} \varphi(s)ds$ then,*

(i) if $\alpha = 0$, then

$$\mathrm{sf}(D_\alpha, D_\alpha + \varphi) = \begin{cases} \lfloor c \rfloor, & c \in \mathbb{Z} \\ \lfloor c \rfloor + \frac{1}{2}\,\mathrm{sgn}(c) - \frac{1}{2} & \textit{otherwise,} \end{cases}$$

(ii) if $\alpha \neq 0$, then

$$\mathrm{sf}(D_\alpha, D_\alpha + \varphi) = \begin{cases} \lfloor c \rfloor, & \{c\} < \alpha \\ \lfloor c \rfloor + \frac{1}{2}\,\mathrm{sgn}(c) + \frac{1}{2}, & \{c\} = \alpha \\ \lfloor c \rfloor + \mathrm{sgn}(c), & \{c\} > \alpha \end{cases}$$

7.2.5 The Anomaly for the Operator $\boldsymbol{D_A}$

Finally, here we compute the anomaly of the operator $\boldsymbol{D_A}$ arising from the Dirac operator D_α on $L_2[0, 2\pi]$. For simplicity, we assume that $\alpha = 0$. By Theorem 6.6.1 the anomaly $\mathrm{Anom}(\boldsymbol{D_A})$ exists, if the limit

$$\lim_{t\to\infty} \frac{1}{t^2} \int_{-t}^{t} \sqrt{t^2 - v^2}\xi(v; D_0 + \varphi, D_0)dv$$

exists. By Theorem 7.2.10 we have $\xi(\cdot; D_0 + \varphi, D_0) = \lfloor c \rfloor + \mathrm{sgn}(c)\sum_{n\in\mathbb{Z}} \chi_{(n,n+\{c\})}$ a.e.. It is clear that

$$\lim_{t\to\infty} \frac{1}{t^2} \int_{-t}^{t} \sqrt{t^2 - v^2}\lfloor c \rfloor dv = \frac{\pi \lfloor c \rfloor}{2}.$$

Proposition 7.2.14 *We have* $\lim_{t\to\infty} \frac{1}{t^2} \int_{-t}^{t} \sqrt{t^2 - v^2}\sum_{n\in\mathbb{Z}} \chi_{(n,n+\{c\})}(v)dv = \frac{\pi\{c\}}{2}$.

Proof We note firstly that

$$\lim_{t\to\infty} \frac{1}{t^2} \int_{-t}^{t} \sqrt{t^2 - v^2}\sum_{n\in\mathbb{Z}} \chi_{(n,n+\{c\})}(v)dv = \lim_{t\to\infty} \frac{1}{t^2} \sum_{n=-t}^{t-1} \int_{n}^{n+\{c\}} \sqrt{t^2 - v^2}dv$$

Using

$$\sqrt{t^2-u^2}-\sqrt{t^2-n^2}=-\frac{u^2-n^2}{\sqrt{t^2-u^2}+\sqrt{t^2+n^2}}$$

we have

$$\lim_{t\to\infty}\frac{1}{t^2}\sum_{n=-t}^{t-1}\int_n^{n+\{c\}}\sqrt{t^2-v^2}dv=\lim_{t\to\infty}\frac{1}{t^2}\sum_{n=-t}^{t-1}\int_n^{n+\{c\}}\sqrt{t^2-n^2}dv$$

$$=\{c\}\lim_{t\to\infty}\frac{1}{t^2}\sum_{n=-t}^{t-1}\sqrt{t^2-n^2}=\{c\}\lim_{t\to\infty}\frac{1}{t}\sum_{n=-t}^{t-1}\sqrt{1-\left(\frac{n}{t}\right)^2}.$$

Since the latter integral is a Riemann sum for the integral $\int_{-1}^{1}\sqrt{1-s^2}ds$, we obtain that

$$\lim_{t\to\infty}\frac{1}{t}\sum_{n=-t}^{t-1}\sqrt{1-\left(\frac{n}{t}\right)^2}=\frac{\pi}{2},$$

which suffices to conclude the proof. □

As an immediate corollary of the above proposition, we obtain the following

Corollary 7.2.15 *Suppose that D_0 is the Dirac operator on $L_2[0,2\pi]$ with periodic boundary conditions and let $\varphi\in C^1[0,2\pi], \varphi(0)=\varphi(2\pi)$. With $c=\frac{1}{2\pi}\int_0^{2\pi}\varphi(s)ds$ the anomaly* $\mathrm{Anom}(\boldsymbol{D_A})$ *of the operator $\boldsymbol{D_A}$ exists and*

$$\mathrm{Anom}(\boldsymbol{D_A})=\lfloor c\rfloor+\mathrm{sgn}(c)\{c\}.$$

Remark 7.2.16 In conclusion, we note the discrepancy between the Witten index $W_s(\boldsymbol{D_A})$ and the anomaly $\mathrm{Anom}(\boldsymbol{D_A})$ for the operator $\boldsymbol{D_A}$ in the case when the endpoints $A_\pm$ are not invertible, so that the operator $\boldsymbol{D_A}$ is not Fredholm. Indeed, as before throughout this section, let D_0 be the Dirac operator on $L_2[0,2\pi]$ with periodic boundary conditions and let $\varphi\in C^1[0,2\pi], \varphi(0)=\varphi(2\pi)$ and $c=\frac{1}{2\pi}\int_0^{2\pi}\varphi(s)ds$. If c is not an integer, then Theorem 7.2.11 implies that

$$W_s(\boldsymbol{D_A})=\lfloor c\rfloor+\frac{1}{2}\mathrm{sgn}(c),$$

while Corollary 7.2.15 guarantees that

$$\mathrm{Anom}(\boldsymbol{D_A})=\lfloor c\rfloor+\mathrm{sgn}(c)\{c\}.$$

Thus, unless $\{c\} = \frac{1}{2}$ we have that

$$W_s(\boldsymbol{D}_A) \neq \mathrm{Anom}(\boldsymbol{D}_A).$$

Furthermore, by Theorem 7.2.13, the anomaly $\mathrm{Anom}(\boldsymbol{D}_A)$ is not equal to $\mathrm{sf}(D_0, D_0 + \varphi)$ either. ◇

References

[ABHN01] W. Arendt, C. Batty, M. Hieber, F. Neubrander, *Vector-Valued Laplace Transforms and Cauchy Problems*. Monographs in Mathematics, vol. 96 (Birkhäuser Verlag, Basel, 2001). MR 1886588

[AC79] Y. Aharonov, A. Casher, Ground state of a spin-$\frac{1}{2}$ charged particle in a two-dimensional magnetic field. Phys. Rev. A (3) **19**(6), 2461–2462 (1979). MR 535300

[ACDS09] N. Azamov, A. Carey, P. Dodds, F. Sukochev, Operator integrals, spectral shift, and spectral flow. Canad. J. Math. **61**(2), 241–263 (2009). MR 2504014

[ACS07] N. Azamov, A. Carey, F. Sukochev, The spectral shift function and spectral flow. Commun. Math. Phys. **276**(1), 51–91 (2007). MR 2342288

[Aib16] D. Aiba, Absence of zero resonances of massless Dirac operators. Hokkaido Math. J. **45**(2), 263–270 (2016). MR 3532132

[APS76] M. Atiyah, V. Patodi, I. Singer, Spectral asymmetry and Riemannian geometry. III. Math. Proc. Cambridge Philos. Soc. **79**, 71–99 (1976)

[AS64] M. Abramowitz, I. Stegun, *Handbook of Mathematical Functions with Formulas, Graphs, and Mathematical Tables*. National Bureau of Standards Applied Mathematics Series, vol. 55, For sale by the Superintendent of Documents (U.S. Government Printing Office, Washington, 1964). MR 0167642

[BBW93] B. Booß Bavnbek, K. Wojciechowski, *Elliptic Boundary Problems for Dirac Operators*. Mathematics: Theory & Applications (Birkhäuser, Boston, 1993). MR 1233386

[BCP+06] M.-T. Benameur, A. Carey, J. Phillips, A. Rennie, F. Sukochev, K. Wojciechowski, *An Analytic Approach to Spectral Flow in von Neumann Algebras*. Analysis, Geometry and Topology of Elliptic Operators (World Science Publication, Hackensack, 2006), pp. 297–352. MR 2246773

[BE11] A. Balinsky, W. Evans, *Spectral Analysis of Relativistic Operators* (Imperial College Press, London, 2011). MR 2779257

[BES08] A. Balinsky, W. Evans, Y. Saito, Dirac-Sobolev inequalities and estimates for the zero modes of massless Dirac operators. J. Math. Phys. **49**(4), 043514 (2008). MR 2412309

[BGG+87] D. Bollé, F. Gesztesy, H. Grosse, W. Schweiger, B. Simon, Witten index, axial anomaly, and Krein's spectral shift function in supersymmetric quantum mechanics. J. Math. Phys. **28**(7), 1512–1525 (1987). MR 894842

[BGGS87] D. Bollé, F. Gesztesy, H. Grosse, B. Simon, Krein's spectral shift function and Fredholm determinants as efficient methods to study supersymmetric quantum mechanics. Lett. Math. Phys. **13**(2), 127–133 (1987). MR 886147

[BGW95] H. Blancarte, B. Grébert, R. Weder, High- and low-energy estimates for the Dirac equation. J. Math. Phys. **36**(3), 991–1015 (1995). MR 1317420

A. Carey, G. Levitina, *Index Theory Beyond the Fredholm Case*, Lecture Notes in Mathematics 2323, https://doi.org/10.1007/978-3-031-19436-8

[Bha96] R. Bhatia, *Matrix Analysis*. Graduate Texts in Mathematics (Springer, New York, 1996)

[BR99] V. Bruneau, D. Robert, Asymptotics of the scattering phase for the Dirac operator: high energy, semi-classical and non-relativistic limits. Ark. Mat. **37**(1), 1–32 (1999). MR 1673424

[BS66] M. Birman, M. Solomyak, *Double Stieltjes Operator Integrals. I*. Problems of Mathematical Physics, No. 1, Spectral Theory, Diffraction Problems (Russian) (Izdat. Leningrad. University, Leningrad, 1966), pp. 33–67

[BS67] M. Birman, M. Solomyak, *Double Stieltjes Operator Integrals. II*. Problems of Mathematical Physics, No. 2, Spectral Theory, Diffraction Problems (Russian) (Izdat. Leningrad. University, Leningrad, 1967), pp. 26–60. MR 0234304

[BS73] M. Birman, M. Solomyak, *Double Stieltjes Operator Integrals. III*. Problems of Mathematical Physics, No. 6 (Russian) (Izdat. Leningrad. University, Leningrad, 1973), pp. 27–53. MR 0348494

[BS87] M. Birman, M. Solomyak, *Spectral Theory of Selfadjoint Operators in Hilbert Space*. Mathematics and its Applications (Soviet Series) (D. Reidel Publishing, Dordrecht, 1987). Translated from the 1980 Russian original by S. Khrushchëv and V. Peller. MR 1192782

[BS96] M. Birman, M. Solomyak, Tensor product of a finite number of spectral measures is always a spectral measure. Integral Equ. Operator Theory **24**(2), 179–187 (1996). MR 1371945

[BS03] M. Birman, M. Solomyak, Double operator integrals in a Hilbert space. Integral Equ. Operator Theory **47**(2), 131–168 (2003). MR 2002663

[Cal78] C. Callias, Axial anomalies and index theorems on open spaces. Commun. Math. Phys. **62**(3), 213–234 (1978). MR 507780

[CGG+16] A. Carey, F. Gesztesy, H. Grosse, G. Levitina, D. Potapov, F. Sukochev, D. Zanin, Trace formulas for a class of non-Fredholm operators: a review. Rev. Math. Phys. **28**(10), 1630002 (2016). MR 3572626

[CGK15] A. Carey, H. Grosse, J. Kaad, Anomalies of Dirac type operators on Euclidean space. Commun. Math. Phys. **335**(1), 445–475 (2015). MR 3314509

[CGK16] A. Carey, H. Grosse, J. Kaad, On a spectral flow formula for the homological index. Adv. Math. **289**, 1106–1156 (2016). MR 3439708

[CGK+18] A. Carey, F. Gesztesy, J. Kaad, G. Levitina, R. Nichols, D. Potapov, F. Sukochev, On the global limiting absorption principle for massless Dirac operators. Ann. Henri Poincaré **19**(7), 1993–2019 (2018). MR 3816195

[CGL+16a] A. Carey, F. Gesztesy, G. Levitina, R. Nichols, D. Potapov, F. Sukochev, Double operator integral methods applied to continuity of spectral shift functions. J. Spectr. Theory **6**(4), 747–779 (2016). MR 3584182

[CGL+16b] A. Carey, F. Gesztesy, G. Levitina, D. Potapov, F. Sukochev, D. Zanin, On index theory for non-Fredholm operators: a $(1 + 1)$-dimensional example. Math. Nachr. **289**(5–6), 575–609 (2016). MR 3486146

[CGL+22] A. Carey, F. Gesztesy, G. Levitina, R. Nichols, F. Sukochev, D. Zanin, *The Limiting Absorption Principle for Massless Dirac Operators, Properties of Spectral Shift Functions, and an Application to the Witten Index of Non-fredholm Operators*. Memoirs of the European Mathematical Society (to appear, 2022)

[CGLS16a] A. Carey, F. Gesztesy, G. Levitina, F. Sukochev, On the index of a non-Fredholm model operator. Oper. Matrices **10**(4), 881–914 (2016). MR 3584682

[CGLS16b] A. Carey, F. Gesztesy, G. Levitina, F. Sukochev, *The Spectral Shift Function and the Witten Index*. Spectral theory and mathematical physics. Operator Theory: Advances and Applications, vol. 254 (Birkhäuser/Springer, Cham, 2016), pp. 71–105. MR 3526446

[CGP+15] A. Carey, V. Gayral, J. Phillips, A. Rennie, F. Sukochev, Spectral flow for nonunital spectral triples. Canad. J. Math. **67**(4), 759–794 (2015). MR 3361012

[CGP+17] A. Carey, F. Gesztesy, D. Potapov, F. Sukochev, Y. Tomilov, On the Witten index in terms of spectral shift functions. J. Anal. Math. **132**, 1–61 (2017). MR 3666804

[CGRS14] A. Carey, V. Gayral, A. Rennie, F.A. Sukochev, Index theory for locally compact noncommutative geometries. Mem. Amer. Math. Soc. **231**(1085), vi+130 (2014). MR 3221983

[CHO82] A. Carey, C. Hurst, D. O'Brien, Automorphisms of the canonical anticommutation relations and index theory. J. Funct. Anal. **48**(3), 360–393 (1982). MR 678177

[CK17] A. Carey, J. Kaad, Topological invariance of the homological index. J. Reine Angew. Math. **729**, 229–261 (2017). MR 3680375

[CLPS22] A. Carey, G. Levitina, D. Potapov, F. Sukochev, The Witten index and the spectral shift function. Rev. Math. Phys. **34**(5), 2250011 (2022). MR 4434952

[CM95] A. Connes, H. Moscovici, The local index formula in noncommutative geometry. Geom. Funct. Anal. GAFA **5**(2), 174–243 (1995)

[CP86] R. Carey, J. Pincus, Index theory for operator ranges and geometric measure theory, in *Geometric Measure Theory and the Calculus of Variations (Arcata, Calif., 1984)*. Proceedings of Symposia in Pure Mathematics, vol. 44 (American Mathematical Society, Providence, 1986), pp. 149–161. MR 840271

[CP98] A. Carey, J. Phillips, Unbounded Fredholm modules and spectral flow. Canad. J. Math. **50**(4), 673–718 (1998). MR 1638603

[CP04] A. Carey, J. Phillips, Spectral flow in Fredholm modules, eta invariants and the JLO cocycle. K-Theory **31**(2), 135–194 (2004). MR 2053481

[CPS09] A. Carey, D. Potapov, F. Sukochev, Spectral flow is the integral of one forms on the Banach manifold of self adjoint Fredholm operators. Adv. Math. **222**(5), 1809–1849 (2009). MR 2555913

[DK56] Y. Daleckiĭ, S. Kreĭn, Integration and differentiation of functions of Hermitian operators and applications to the theory of perturbations. Voronež. Gos. Univ. Trudy Sem. Funkcional. Anal. **1956**(1), 81–105 (1956). MR 0084745

[dPWS02] B. de Pagter, H. Witvliet, F. Sukochev, Double operator integrals. J. Funct. Anal. **192**(1), 52–111 (2002). MR 1918492

[Dug66] J. Dugundji, *Topology* (Allyn and Bacon, Boston, 1966). MR 0193606

[DZ98] X. Dai, W. Zhang, Higher spectral flow. J. Funct. Anal. **157**(2), 432–469 (1998). MR 1638328

[Geo13] M. Georgescu, Spectral flow in semifinite von neumann algebras, Ph.D. Thesis, Department of Mathematics and Statistics, University of Victoria (2013)

[Get93] E. Getzler, The odd Chern character in cyclic homology and spectral flow. Topology **32**(3), 489–507 (1993). MR 1231957

[GK69] I. Gohberg, M. Krein, *Introduction to the Theory of Linear Nonselfadjoint Operators*. Translated from the Russian by A. Feinstein. Translations of Mathematical Monographs, vol. 18 (American Mathematical Society, Providence, 1969). MR 0246142

[GLM+11] F. Gesztesy, Y. Latushkin, K. Makarov, F. Sukochev, Y. Tomilov, The index formula and the spectral shift function for relatively trace class perturbations. Adv. Math. **227**(1), 319–420 (2011). MR 2782197

[GLST15] F. Gesztesy, Y. Latushkin, F. Sukochev, Y. Tomilov, Some operator bounds employing complex interpolation revisited, in *Operator Semigroups Meet Complex Analysis, Harmonic Analysis and Mathematical Physics*. Operator Theory: Advances and Applications, vol. 250 (Birkhäuser/Springer, Cham, 2015), pp. 213–239. MR 3468218

[GS88] F. Gesztesy, B. Simon, Topological invariance of the Witten index. J. Funct. Anal. **79**(1), 91–102 (1988). MR 950085

[HJ12] R. Horn, C. Johnson, *Matrix Analysis*, 2nd edn. (Cambridge University Press, New York, 2012)

[JN01] A. Jensen, G. Nenciu, A unified approach to resolvent expansions at thresholds. Rev. Math. Phys. **13**(6), 717–754 (2001). MR 1841744

[KOY15] H. Kalf, T. Okaji, O. Yamada, The Dirac operator with mass $m_0 \geq 0$: non-existence of zero modes and of threshold eigenvalues. Doc. Math. **20**, 37–64 (2015). MR 3398708

[Kre53] M. Krein, On the trace formula in perturbation theory. Mat. Sbornik N.S. **33**(75), 597–626 (1953). MR 0060742
[Kre62] M. Krein, On perturbation determinants and a trace formula for unitary and self-adjoint operators. Dokl. Akad. Nauk SSSR **144**, 268–271 (1962). MR 0139006
[Kre81] V. Kreĭn, M. Yavryan, Spectral shift functions that arise in perturbations of a positive operator. J. Operator Theory **6**(1), 155–191 (1981). MR 637009
[L34] K. Löwner, Über monotone Matrixfunktionen. Math. Z. **38**(1), 177–216 (1934). MR 1545446
[Les05] M. Lesch, The uniqueness of the spectral flow on spaces of unbounded self-adjoint Fredholm operators, in *Spectral Geometry of Manifolds with Boundary and Decomposition of Manifolds*. Contemporary Mathematics, vol. 366 (American Mathematical Society, Providence, 2005), pp. 193–224. MR 2114489
[LL01] E. Lieb, M. Loss, *Analysis*. Graduate Studies in Mathematics, vol. 14, 2nd edn. (American Mathematical Society, Providence, 2001). MR 1817225
[Lod98] J.-L. Loday, *Cyclic Homology*. Grundlehren der mathematischen Wissenschaften [Fundamental Principles of Mathematical Sciences], vol. 301, 2nd edn. (Springer, Berlin, 1998). Appendix E by María O. Ronco, Chapter 13 by the author in collaboration with Teimuraz Pirashvili. MR 1600246
[LSVZ18] G. Levitina, F. Sukochev, D. Vella, D. Zanin, Schatten class estimates for the Riesz map of massless Dirac operators. Integral Equ. Operator Theory **90**(2), 19, 36 (2018). MR 3798015
[LSZ20] G. Levitina, F. Sukochev, D. Zanin, Cwikel estimates revisited. Proc. Lond. Math. Soc. (3) **120**(2), 265–304 (2020). MR 4008371
[Mel93] R. Melrose, *The Atiyah-Patodi-Singer Index Theorem*. Research Notes in Mathematics, vol. 4 (A K Peters, Ltd., Wellesley, 1993). MR 1348401
[Mic89] J. Mickelsson, *Current Algebras and Groups*. Plenum Monographs in Nonlinear Physics (Plenum Press, New York, 1989). MR 1032521
[MP97] R. Melrose, P. Piazza, An index theorem for families of Dirac operators on odd-dimensional manifolds with boundary. J. Differ. Geom. **46**(2), 287–334 (1997). MR 1484046
[Pal91] B. Palka, *An Introduction to Complex Function Theory*. Undergraduate Texts in Mathematics (Springer, New York, 1991). MR 1078017
[Pel85] V. Peller, Hankel operators in the theory of perturbations of unitary and selfadjoint operators. Funktsional. Anal. i Prilozhen. **19**(2), 37–51, 96 (1985). MR 800919
[Pel90] V. Peller, Hankel operators in the perturbation theory of unbounded selfadjoint operators. *Analysis and Partial Differential Equations*. Lecture Notes in Pure and Applied Mathematics, vol. 122 (Dekker, New York, 1990), pp. 529–544. MR 1044807
[Pel16] V. Peller, Multiple operator integrals in perturbation theory. Bullet. Math. Sci. **6**(1), 15–88 (2016)
[Phi96] J. Phillips, Self-adjoint Fredholm operators and spectral flow. Canad. Math. Bull. **39**(4), 460–467 (1996). MR 1426691
[PS09] D. Potapov, F. Sukochev, Unbounded Fredholm modules and double operator integrals. J. Reine Angew. Math. **626**, 159–185 (2009). MR 2492993
[PS10] D. Potapov, F. Sukochev, Double operator integrals and submajorization. Math. Model. Nat. Phenom. **5**(4), 317–339 (2010). MR 2662461
[PS11] D. Potapov, F. Sukochev, Operator-Lipschitz functions in Schatten-von Neumann classes. Acta Math. **207**(2), 375–389 (2011). MR 2892613
[PSS13] D. Potapov, A. Skripka, F. Sukochev, Spectral shift function of higher order. Invent. Math. **193**(3), 501–538 (2013). MR 3091975
[Pus08] A. Pushnitski, The spectral flow, the Fredholm index, and the spectral shift function. *Spectral Theory of Differential Operators*. American Mathematical Society Translations: Series 2, vol. 225 (American Mathematical Society, Providence, 2008), pp. 141–155. MR 2509781

[RS80] M. Reed, B. Simon, *Methods of Modern Mathematical Physics. I*. Functional Analysis, 2nd edn. (Academic Press/Harcourt Brace Jovanovich, Publishers, New York, 1980). MR 751959

[RS95] J. Robbin, D. Salamon, The spectral flow and the Maslov index. Bull. Lond. Math. Soc. **27**(1), 1–33 (1995). MR 1331677

[Saf01] O. Safronov, Spectral shift function in the large coupling constant limit. J. Funct. Anal. **182**(1), 151–169 (2001). MR 1829245

[Sch59] J. Schwinger, Field theory commutators. Phys. Rev. Lett. **3**, 296–297 (1959)

[Sim98] B. Simon, Spectral averaging and the Krein spectral shift. Proc. Am. Math. Soc. **126**(5), 1409–1413 (1998). MR 1443857

[Sim05] B. Simon, *Trace Ideals and Their Applications*. Mathematical Surveys and Monographs, vol. 120, 2nd edn. (American Mathematical Society, Providence, 2005). MR 2154153

[ST19] A. Skripka, A. Tomskova, *Multilinear Operator Integrals*. Lecture Notes in Mathematics, vol. 2250 (Springer, Cham, 2019)©2019. Theory and applications. MR 3971571

[Sto90] M. Stone, *Linear Transformations in Hilbert Space*. American Mathematical Society Colloquium Publications, vol. 15 (American Mathematical Society, Providence, 1990). Reprint of the 1932 original. MR 1451877

[Sto96] M. Stone, Spectral flow, Magnus force, and mutual friction via the geometric optics limit of Andreev reflection. Phys. Rev. B **54**, 13222–13229 (1996)

[SU08a] Y. Saito, T. Umeda, The asymptotic limits of zero modes of massless Dirac operators. Lett. Math. Phys. **83**(1), 97–106 (2008). MR 2377949

[SU08b] Y. Saito, T. Umeda, The zero modes and zero resonances of massless Dirac operators. Hokkaido Math. J. **37**(2), 363–388 (2008). MR 2415906

[TdA11] R. Tiedra de Aldecoa, Asymptotics near $\pm m$ of the spectral shift function for Dirac operators with non-constant magnetic fields. Commun. Partial Differ. Equ. **36**(1), 10–41 (2011). MR 2763346

[Wei80] J. Weidmann, *Linear Operators in Hilbert Spaces*. Graduate Texts in Mathematics, vol. 68 (Springer, New York, 1980). Translated from the German by Joseph Szücs. MR 566954

[Wid41] D. Widder, *The Laplace Transform*. Princeton Mathematical Series, vol. 6 (Princeton University Press, Princeton, 1941). MR 0005923

[Wit82] E. Witten, Constraints on supersymmetry breaking. Nuclear Phys. B **202**(2), 253–316 (1982). MR 668987

[Yaf92] D. Yafaev, *Mathematical Scattering Theory*. Translations of Mathematical Monographs, vol. 105 (American Mathematical Society, Providence, 1992). General theory, Translated from the Russian by J. R. Schulenberger. MR 1180965

[Yaf05] D. Yafaev, A trace formula for the Dirac operator. Bull. Lond. Math. Soc. **37**(6), 908–918 (2005). MR 2186724

[ZG13] Y. Zhong, G.L. Gao, Some new results about the massless Dirac operator. J. Math. Phys. **54**(4), 043510 (2013). MR 3088812

LECTURE NOTES IN MATHEMATICS

Editors in Chief: J.-M. Morel, B. Teissier;

Editorial Policy

1. Lecture Notes aim to report new developments in all areas of mathematics and their applications – quickly, informally and at a high level. Mathematical texts analysing new developments in modelling and numerical simulation are welcome.

   Manuscripts should be reasonably self-contained and rounded off. Thus they may, and often will, present not only results of the author but also related work by other people. They may be based on specialised lecture courses. Furthermore, the manuscripts should provide sufficient motivation, examples and applications. This clearly distinguishes Lecture Notes from journal articles or technical reports which normally are very concise. Articles intended for a journal but too long to be accepted by most journals, usually do not have this "lecture notes" character. For similar reasons it is unusual for doctoral theses to be accepted for the Lecture Notes series, though habilitation theses may be appropriate.

2. Besides monographs, multi-author manuscripts resulting from SUMMER SCHOOLS or similar INTENSIVE COURSES are welcome, provided their objective was held to present an active mathematical topic to an audience at the beginning or intermediate graduate level (a list of participants should be provided).

   The resulting manuscript should not be just a collection of course notes, but should require advance planning and coordination among the main lecturers. The subject matter should dictate the structure of the book. This structure should be motivated and explained in a scientific introduction, and the notation, references, index and formulation of results should be, if possible, unified by the editors. Each contribution should have an abstract and an introduction referring to the other contributions. In other words, more preparatory work must go into a multi-authored volume than simply assembling a disparate collection of papers, communicated at the event.

3. Manuscripts should be submitted either online at www.editorialmanager.com/lnm to Springer's mathematics editorial in Heidelberg, or electronically to one of the series editors. Authors should be aware that incomplete or insufficiently close-to-final manuscripts almost always result in longer refereeing times and nevertheless unclear referees' recommendations, making further refereeing of a final draft necessary. The strict minimum amount of material that will be considered should include a detailed outline describing the planned contents of each chapter, a bibliography and several sample chapters. Parallel submission of a manuscript to another publisher while under consideration for LNM is not acceptable and can lead to rejection.

4. In general, **monographs** will be sent out to at least 2 external referees for evaluation.

   A final decision to publish can be made only on the basis of the complete manuscript, however a refereeing process leading to a preliminary decision can be based on a pre-final or incomplete manuscript.

   Volume Editors of **multi-author works** are expected to arrange for the refereeing, to the usual scientific standards, of the individual contributions. If the resulting reports can be forwarded to the LNM Editorial Board, this is very helpful. If no reports are forwarded or if other questions remain unclear in respect of homogeneity etc, the series editors may wish to consult external referees for an overall evaluation of the volume.

5. Manuscripts should in general be submitted in English. Final manuscripts should contain at least 100 pages of mathematical text and should always include

   - a table of contents;
   - an informative introduction, with adequate motivation and perhaps some historical remarks: it should be accessible to a reader not intimately familiar with the topic treated;
   - a subject index: as a rule this is genuinely helpful for the reader.
   - For evaluation purposes, manuscripts should be submitted as pdf files.

6. Careful preparation of the manuscripts will help keep production time short besides ensuring satisfactory appearance of the finished book in print and online. After acceptance of the manuscript authors will be asked to prepare the final LaTeX source files (see LaTeX templates online: https://www.springer.com/gb/authors-editors/book-authors-editors/manuscriptpreparation/5636) plus the corresponding pdf- or zipped ps-file. The LaTeX source files are essential for producing the full-text online version of the book, see http://link.springer.com/bookseries/304 for the existing online volumes of LNM). The technical production of a Lecture Notes volume takes approximately 12 weeks. Additional instructions, if necessary, are available on request from lnm@springer.com.

7. Authors receive a total of 30 free copies of their volume and free access to their book on SpringerLink, but no royalties. They are entitled to a discount of 33.3 % on the price of Springer books purchased for their personal use, if ordering directly from Springer.

8. Commitment to publish is made by a *Publishing Agreement*; contributing authors of multiauthor books are requested to sign a *Consent to Publish form*. Springer-Verlag registers the copyright for each volume. Authors are free to reuse material contained in their LNM volumes in later publications: a brief written (or e-mail) request for formal permission is sufficient.

Addresses:

Professor Jean-Michel Morel, CMLA, École Normale Supérieure de Cachan, France
E-mail: moreljeanmichel@gmail.com

Professor Bernard Teissier, Equipe Géométrie et Dynamique,
Institut de Mathématiques de Jussieu – Paris Rive Gauche, Paris, France
E-mail: bernard.teissier@imj-prg.fr

Springer: Ute McCrory, Mathematics, Heidelberg, Germany,
E-mail: lnm@springer.com

Printed in the United States
by Baker & Taylor Publisher Services

Printed in the United States
by Baker & Taylor Publisher Services